Estuarine Dynamics in Flux: Time to Deepen Our Understanding

Shiva

Table of Contents

Introduction

The multifaceted relevance of estuaries warrants further studies on current estuarine dynamics and how these systems are changing. Within the past decade, there has been increasing research of estuarine areas within global studies on biogeochemical cycles of carbon and nitrogen in efforts to understand the flux of terrestrial sediment from the land to the sea (Syvitski and Kettner et al., 2011; Bauer et al., 2013; Battin et al., 2008; St-Onge and Hillaire-Marcel, 2001). Historical changes in sea level rise, river output, and earthquake activity can be visible within sediment cores retrieved from estuarine areas that provide insight into environmental changes and past events (Hamilton et al., 2015; Baek et al., 2017; Bianchi and Allison, 2009). Locally, estuarine morphodynamics also impacts coastal infrastructure through erosion and deposition (Dugan et al., 2011). Estuarine sediment dynamics also impact fisheries and ecology as many juvenile fish species depend on nutrient flux into estuarine areas (Carr-Harris et al., 2015). The amount, grain size, and movement of sediment within an estuary affect light penetration, which is critical for aquatic vegetation that forms the base of marine food chains (Moore et al., 1997; Levin et al., 2000).

Most of the world has a bedrock coastline, yet rocky coastline areas are the most understudied within the field of coastal studies (Syvitski et al., 2005). As will be described further below, abundant literature is available describing and classifying the morphodynamics of unconfined, softer bank, estuaries[1]. Such estuaries have been classified into different facies (Dalrymple et al., 1992; Dalrymple and Choi, 2007; Liangqing & Galloway, 1991), successfully modelled (Skinner et al, 2015; Hibma et al., 2003; Hu et al., 2009) and their morphodynamics are well understood (Uncles, 1991; Isla & Bujalesky, 2004; Dalrymple et al., 2012). Although many studies have been conducted on the morphology, sedimentation, and processes within bedrock fjords (Syvitski et al., 2012; Shaw et al., 2017; Svendsen et al., 2002; St-Onge and Hillaire-Marcel, 2001), the impact of bedrock confinement on estuarine morphodynamics in non-fjordic or a-typical fjordic areas is comparatively understudied. Within the bedrock confined Skeena Estuary in northwestern Canada, empirical observations are not matching with typical estuarine and delta classification schemes and are producing conflicting modelling result

---

1 Unconfined estuaries and deltas are those that can adjust their topography due to relatively erodible banks and transport patterns that are not disrupted by multiple large and immobile obstructions.

(McLaren, 2016). This research will provide comprehensive insight into the morphodynamics of the Skeena River Estuary because of its bedrock structure and understudied nature.

Literature Review on Estuaries

The following section addresses estuaries in literature and theory to provide a general contextual understanding of estuaries relevant to the book. Although a general overview of estuarine processes and features is provided below, the following literature and descriptions will be curtailed to estuarine phenomena relevant to the Skeena River and Estuary study site. Thus, certain phenomena observed in estuaries drastically different from the Skeena in terms of processes and physiography, such as intermittently closed, bar-built, and wave dominant estuaries, will not be discussed in detail. Furthermore, the specific conditions and known literature on the Skeena Estuary will be discussed later in the Skeena background section.

Estuary Definition

Typically, estuaries classify as transgressive, semi-enclosed areas that receive sediment from fluvial and marine sources and form where rivers meet the sea (Dalrymple et al., 1992; Wolanski and Elliott, 2007; Bierman and Montgomery, 2014). Estuaries respond to sea-level change by infilling, prograding, and retrograding sediment to create a more gradual transition between the land and sea along shorelines that migrate over time (Flemming, 2011; Wolanksi and Elliott, 2007). As described in Paturej (2008), throughout the 20th-century estuaries have been described in many different ways over time based on their circulation and salinity (Pritchard, 1952; Prichard, 1955; Hansen and Rattray, 1966; Duxbury et al., 2002), physiography (Pritchard, 1967; Perillo, 1995), and dominant energy and sediment transport processes (Stommel, 1951; Darlymple et al., 1992). In terms of circulation, estuaries can be hyperhaline, highly stratified, partially, or well mixed (Pritchard 1952; Pritchard,1955; Day, 1980). This can further break into salt-wedge, partially mixed, well-mixed and fjord-like estuaries (Hansen and Rattray, 1966; Duxbury et al., 2002). Regarding physiography, estuaries classify as coastal plain (drowned valleys), tectonically produced, bar-built, fjord-type, or formed in river deltas (Pritchard, 1967). Some classifications divide this further into primary and secondary morphologies (Perillo, 1995). Finally, based on the influence of specific processes, estuaries have been described as intermittently closed, wave, or tidally dominant (Stommel, 1951; Dalrymple et al., 1992). Original classifications (such as Pritchard (1967)) have been modified, adapted, and described by many sources to this day. Due to difficulties in digital access of older sources and their wide adaption, some of the estuarine classification schemes presented here may originate from older sources but are cited in more current work.

Estuarine Processes

Waves, Tides, and Rivers

Rivers, waves, and tides induce varied stratification, circulation, and sediment transport within an estuary depending on the conditions. Macrotidal is a term used to describe estuaries that are tidally dominated and with large water level variations between tides (Hayes, 1975; Boothroyd, 1978; Bierman and Montgomery, 2014). Estuaries with large tidal amplitudes (>4m) are referred to as macrotidal, followed by mesotidal (water level change of 2-4m), and microtidal (change <2m) (Boothroyd, 1978). In traditional systems, microtidal estuaries are largely wave-dominated, and mesotidal estuaries are considered mixed wave and tidal energy (Boothroyd, 1978). If oceanic energy can match the fluvial energy to create the mixing required for an estuary, an estuary can still form in areas receiving high fluvial inputs. However, once fluvial deposition dominates and prograde seaward, this is called a delta rather than an estuary.

Generally, local morphology and bathymetry can impact tide and wave propagation. Waves have an amplified effect in low water conditions, whereas tides can be amplified from a converging bank morphology (e.g., funnel) (Friedman and Sanders, 1978). On a larger, more broad scale, offshore winds and storm dissipation can impact wave formation whereas latitude, seasons, and position of the moon and sun can impact tidal amplitudes. The local climate, drainage area, glaciers, groundwater, vegetation, ground cover and lithology, and other elements of the water cycle such as evaporation can impact fluvial inflows (*Kettner and Syvitski, 2008*).

Coriolis and Centrifugal Force

Coriolis force can also impact fluvial and marine water mixing and sediment transport in large, low-velocity estuarine basins. Georgas and Blumberg (2004) observed that Coriolis acceleration may shift direction with the incoming/outgoing currents in estuarine channels wider than 5 km and is the most evident during slack tides when tidal velocities are lowest. On a study on the Yangtze River mouth, Li et al. (2011) noted that Coriolis force can drag ebb flows northward (a diversion to the right) in the northern hemisphere and southward (a diversion to the left) in the southern hemisphere and affect sediment transport at the estuarine scale. Coriolis initiated diversion of water flow at the river mouth contributed to the formation of a slack water setting that encouraged siltation[2] during certain tides (Li et al., 2011). In a study in the Elbe estuary, Coriolis force leads to inconsistent inwards and outward flows at the river mouth that, along with other factors such as dredging, encouraged siltation (Li et al., 2014). More specifically, Coriolis force in combination with tidal and riverine forces in the Elbe contributed to a shift in the river thalweg approaching the receiving basin, a dextral extension of fluvial material, and the formation of a northwestern tidal flat (Li et al., 2014).

Sea level Rise

Estuaries respond to sea-level rise by infilling submerged river valley, fjord, and basin areas with oceanic and fluvial sediments to produce a more gradual plain between the land, river, and seabed (Wolanski & Elliott, 2007; Hutton & Syvitski, 2008). Estuaries can infill when fluvial inputs are high by forming a delta or from deposits of marine, and glacial sediments at the seabed brought landward by the tides that are not flushed away by river discharge (Friedman and Sanders, 1978). If sea level drops, the river will erode the past the estuarine sediments, now directly at its mouth, to base level and shift to a delta or estuary system that pushes sediment further out to sea (Hutton & Syvitski, 2008).

Storms and Floods

Storms and floods can also play a large role in redistributing sediment, altering morphology, as well as recirculating and oxidizing estuarine waters (Schumann, 2015). A study of the Keurbooms wave dominated, micro-tidal estuary along the coast of South Africa has shown that floods and storms can alter the position of the estuary mouth by creating a break in a protective barrier dune whereby the estuary is exposed to ocean waters (Schumann, 2015). Along with raising the water level, depending on the estuary, high energy events can cause erosion of the estuary banks, alter the position of spits and the estuary mouth, and reroute back barrier channels (Schumann, 2015). Within the Keurbooms estuary, a combination of local morphology and flood or storm magnitude appear to play a large role in determining the location of erosional and depositional activity within the estuary (Schumann, 2015). The impact of storms and floods in a meso or macrotidal estuary can also redistribute harder to entrain sediment, due to a higher transport energy, and be visible in the sedimentary record. However, in meso and macrotidal

---

2 High suspended silt content

estuaries, tidal processes would continue to redistribute and rework most sediment after storm events. Meanwhile, in a microtidal estuary, a storm depositional location may be inaccessible after the water level drops. Thus, the impact of storms and floods would be more evident in low-tidal energy, microtidal areas. Floods can also increase incoming freshwater to the estuary, thereby altering the density gradients[3], circulation, stratification, salt flux, and sediment flux within the estuary.

Estuarine Phenomenon

Estuarine Stratification and Flow

Tides transport sea water landward and seaward and interact with fresh water to form stratified to vertically homogenous conditions with unidirectional or bidirectional flow reversals with depth throughout the water column. Different types of circulation and flow induce varied estuarine circulation and salt propagation (Hansen and Rattray, 1966). The ratio of tidal to river strength induce varied stratification within an estuary (Valle-Levinson, 2011; Geyer, 2010). However, local features of bathymetry and geometry can also play a role (Georgas and Blumberg, 2004). Aside from the strength of the tidal compared to fluvial inflow, the specific water properties such as density, temperature, and salinity can affect stratification (Ritter et al., 2002). Also, stratification and flow can vary over a tidal cycle (Becker et al., 2018) and tidal mean classification schemes do not capture these details. Overall, there are many different equations and techniques used to classify the stratification and circulation of an estuary. Only a few significant terms and phenomena are described below. Specific schemes used to classify the Skeena estuary within this book will be described in further detail within the methods sections.

Stratification in estuaries is typically described at hyperhaline, highly stratified, partially mixed, or well mixed (Pritchard 1952; Pritchard, 1955; Valle-Levinson, 2011; Geyer, 2010). Sometimes well-mixed and vertically homogenous conditions are differentiated. Hyperhaline conditions are relatively rare and occur when evaporation and the enclosed nature of certain basins produce estuarine conditions that are more saline than the surrounding ocean water (Paturej, 2008). Typically, salinity and density are used to assess stratification. In highly stratified estuaries, salinity will be high in deeper waters and low at the surface with a sharp

---

3 For example, higher flood flows have been known to induce turbidity currents (Hizett et al., 2017).

halocline gradient in between (Geyer and Chant, 2006). Partially mixed conditions display a gradual gradient between often less extreme differences in salinity with depth. Well mixed conditions are closer to a vertically homogenous profile or a very slight, gradual increase in salinity with depth (Geyer and Chant, 2006).

Based on the flow and stratification with depth, estuaries classify as salt wedge systems, highly stratified fjords, partially mixed bidirectional flowing, or well-mixed unidirectional flowing estuaries (Hansen and Rattray, 1966; Duxbury et al. 2002). Shown in Figure LIT1, a salt wedge system is highly stratified whereby the denser seawater forces beneath the freshwater at the surface (Valle-Levinson, 2011; Duxbury et al., 2002; Hansen and Rattray, 1966). During the flood tide, seawater flows in the opposite direction at the bottom than the outflowing freshwater at the surface. Highly stratified fjords are similar, but they are distinct in that it is there deep nature that allows for a stark contrast in density and salinity with depth (Figure LIT1). Partially stratified conditions display vertical mixing within the water column causing a reduction in the stark contrast at the surface at with depth (Figure LIT1). Highly and partially stratified estuaries can experience some bidirectional flow reversals with depth depending on the tide. Well mixed conditions are unidirectional throughout the water column as the high mixing does not allow for a flow reversal at depth (Valle-Levinson, 2011; Duxbury et al., 2002; Hansen and Rattray, 1966).

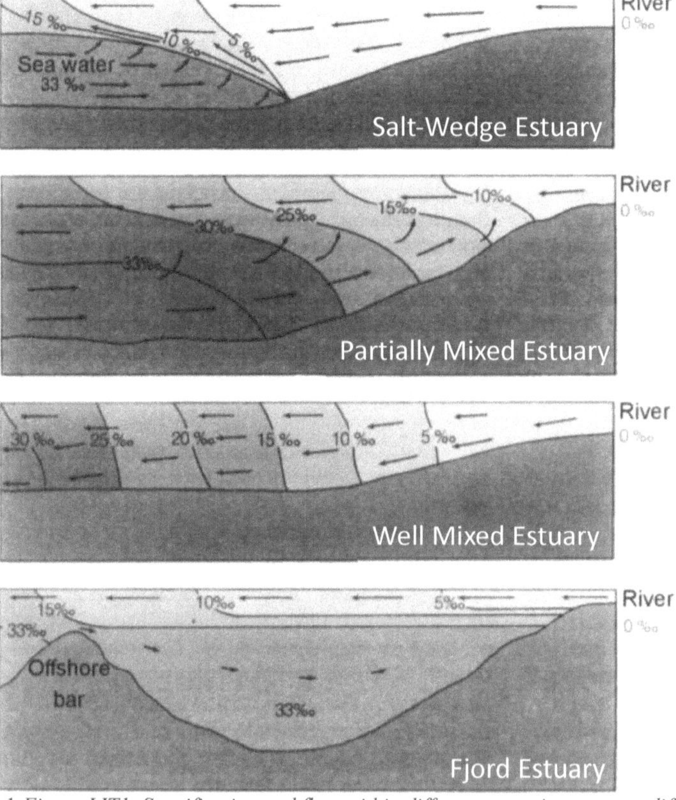

1 *Figure LIT1: Stratification and flow within different estuarine types modified from Duxbury et al. (2002).*

Estuaries are areas where fresh and saltwater meet, but the nature of this interaction can vary drastically and change how salt in propagates through the estuary. When stratification is high, and flow is bidirectional, upstream salt flux propagates through gravitational convection in two-layer flow (Hansen and Rattray, 1966). When the estuary is unidirectional and well-mixed, the upstream salt flux is through diffusion, and gravitational convection is null (Hansen and Rattray, 1966). In deep fjordic areas, advection can be the dominant form of salt flux as the salinity gradient and circulation do not extend to the deeper bottom (Hansen and Rattray, 1966). Due to horizontal advection, stratification increases producing higher salinity in deeper water and a sharp decrease at the surface (Geyer and Chant, 2006). Vertical mixing partially counteracts estuarine stratification and circulation.

Increased tidal currents and mixing over the changing tide can increase the overall vertical mixing of an estuary and decrease estuarine circulation and stratification (Geyer and Chant, 2006). Thus, in theory, higher river inflows with lower tidal ranges should increase estuarine circulation and stratification so long as the freshwater is not entirely dominating the flow. With a more significant river inflow, higher circulation is required to produce the same mean salinity landward than in a well-mixed estuary (Geyer and Chant, 2006). According to Valle-Levinson (2011), the closer the calculated annual mean tidal prism[4] to the mean river inflow over the tidal cycle, the higher the stratification. If the tidal or river inflow is magnitudes greater than another, the higher order of magnitude process will dominate creating a well-mixed water column (Valle-Levinson, 2011). Thus, the calculation of tidal prisms and river inflow over a tidal cycle is one method to assess the mean estimated stratification condition of an estuary. In general, higher stratification induces higher circulation within the estuary and can be caused by a reduction in the tidal range relative to the fluvial input.

Cameron and Pritchard (1963) and Cavalcante (2016) indicate that horizontal bidirectional flow can develop in wide and shallow channels under well mixed conditions and may not indicate a highly stratified estuary and gravitational circulation in the same way as vertical bidirectional flow. Furthermore, Georgas and Blumberg (2004) state that cross-channel horizontal circulation of flow can be induced by channel curvature and varied cross-channel bathymetry. Thus, in assessing horizontal circulation and stratification is not commonly calculated within circulation and stratification parameters used to overall well-mixed to high stratified estuarine conditions (Valle-Levinson, 2011; Hansen and Rattray, 1966).

Estuarine Sediment Dynamics

Multiple drivers in opposite directions such as the river discharge, tidal strength, and wave-wind energy combine to produce the transport energy for a given sediment grain (Dalrymple et al., 1992). What is brought in from the fluvial and marine environment, past deposits on the seabed (e.g.: glacial), the surrounding lithology, and the available energy to break

---

4 Essentially, the volume of incoming tidal seawater

down or rework sediments predominantly determine the available material within an estuary. Some estuaries have a zone of elevated sediment concentrations referred to as the estuarine turbidity maximum (ETM) (Jay et al., 2015). ETM zones are the most common in the coastal plain, salt wedge, and river dominated estuaries whereas high riverine inputs disperse, they produce a zone within the estuary of higher turbidity (Jay et al., 2015). However, ETM zones can also form from turbulent resuspension of material and a lack of flocculation[5]. Site specific geomorphic features such as the position of the river mouth regarding the sea level, the gradient and shape of the bed, the topography of the estuary, and size of the sediment grains influence the way that the available sediments and transport energy interact (Wolanksi and Elliott, 2007). Furthermore, the greater the variation in topography, bed and bank morphology, and inputs to the system, then the greater the variance in the sediment dynamics of the estuary or delta system.

Waves, Tides, Rivers, and Sediment Transport

The net direction of sediment transport will vary locally depending on the force of the tides, waves, and river discharge. Rivers provide sediment and terrestrial material to the estuary while tides and waves induce mixing, resuspension, and marine sediments. Kostaschuk et al. (1989) describe how seasonal fluctuations in river discharge induce a lagged change in height and length of bedforms transporting sediment along the bed. The transport of river plumes can also shift depending on the season based on density, salinity, amount of sediment, and mixing with estuarine waters (Ritter et al., 2002). High river discharge events along with strong ebb tides push sediment seaward (Ritter et al., 2002). A strong flood tide, also referred to as tidal pumping, can transport sediment landward throughout the estuary (Ritter et al., 2002). Friedman and Sanders (1975) described that when tides predominate, fine sediments coat the shore and coarser sediments occupy the channel center. Meanwhile, waves resuspend material and cleanse intertidal sand flats of fine sediment by resuspending fine material that gets carried seaward by the tide (Green, 2011). Intense waves can break apart and re-entrain sediments creating sediments that have been highly reworked.

Over a tidal cycle and different stratification conditions, there can be high local variance in the resuspension and transport of sediment. For example, at the Jemgum study site within the

---

5 The clustering of finer (silt and clay) suspended particles into a larger mass within the water column (Wolanski & Elliott, 2007).

Ems, Becker et al., (2018) describe that when the velocity is lower during the late flood phase or ebb flow, the water column exhibits a two-layer structure where a fluid mud layer at the bed stratifies from the upper water column. At the peak of the Ebb, sediments settle, creating a reduced fluid mud layer. Becker et al. (2018) observed the fluid mud layer at the bed being initially advected upstream and upwards directly after the ebb tide, during the early flood phase. Under rising tides, Kostaschuk and Luternauer (1989) describe a decline in suspended bed-material as the flow becomes stratified during the flood and forms a salt-wedge within the Fraser River Estuary. Under non-stratified conditions, there is a constant exchange of sediment between the bed and the water column within the estuary (Kostaschuk and Luternauer, 1989). During falling tides, the salt-wedge tip entrains sediment that increases the concentration at the wedge head and falls out of transport during lower velocities over the changing tide (Kostaschuk and Luternauer, 1989). However, the specific dynamics of any given estuary will vary depending on the sediment inputs, physiography, and dynamics such as tidal strength.

Sediment Transport of Coarse vs Fine Sediment

Based on the descriptions by Wolanski and Elliott (2007), sediment dynamics will vary drastically depending on the presence of sand or mud. Generally, sand is heavier and non-cohesive in comparison to mud and typically carried in the bedload[6] of the estuary (Wolanski & Elliott, 2007). Sand movement along the bed is described as creeping and saltation by Wolanski & Elliott (2007). Since, sand particles travel on the bed and are larger, estuaries dominated by sand allow for more light penetration into the water column and less absorption of nutrients and heavy minerals. Sand can form dunes and ripples on the bed that can increase bed roughness, interact with waves, form turbulent eddies, and create patchy deposits that may shift with high currents (Friedman and Sanders, 1978). Thus, the formation of dunes on the seabed can encourage water mixing due to their impact on increasing bed roughness (Wolanski & Elliott, 2007).

Mud is cohesive and composed of lightweight silt or clay allowing it to remain in the suspended load for longer. Silt is less cohesive than clay and is more complicated in determining

---

6 Bedload refers to the material that bounces, tumbles, and travels along the bed of the river (Ritter *et al.*, 2011). Bedload generally consists of more coarse grain material than the suspended load (Ritter *et al.*, 2011).

its behavior and effect on biota (Wolanski & Elliott, 2007). According to Wolanski & Elliott (2007) and Friedman and Sanders (1978), highly cohesive clay particles can absorb nutrients and bind with heavy minerals. In high amounts, this can limit nutrient access in the water column for biota and decrease light penetration for photosynthesis. Once deposited, clay can store heavy minerals for long periods that get released when the mud is disturbed (Schoellhamer, 2002). The suspended mud layer is typically thicker closer to the bed and will vary depending on the bioturbation and mixing of the water column occurring at the seabed (Wolanski & Elliott, 2007). Thick mud layers can suffocate biota in high amounts or provide a penetrable home when mixed with some loose, large grain sediments (Friedman and Sanders, 1978). In semi-turbid waters where some light penetration is still possible, biota form mucus membranes called transparent exopolymer particles that bind with cohesive sediments to form aggregates of mud "snow" in the water column called flocs (Wolanski & Elliott, 2007). These flocs allow for more light penetration as opposed to a bath of fine particles that block out light. Biota and wave action can loosen and re-entrain mud sediments and break apart flocs relatively easily if recapping by new sediment does not occur. In summary, the amount of cohesive clay in an estuary can affect light penetration in an estuary. This will affect the organic deposition to the seabed as well as bioturbation within sediments that can have feedbacks on sediment transport and turbidity within the estuary as well as nutrient cycling (Wolanski & Elliott, 2007).

Grain size Trends and Variance in Deposition

Hjulström (1935) developed a curve to describe the relationship between river velocity and the transport, erosion, or deposition of a given grain size type (Ritter et al., 2002). Whereby, channel width, depth, discharge, slope, and boundary conditions (smoothness or roughness of the bed and banks) are the controls on river velocity based on the equations of Chezy (1776) and Manning (1890) as well as Leopold and Maddock (1953) described in Ritter et al. (2002). Discussed by Ritter et al. (2002), the Hjulström's curve relates transport energy to grain size transport considering the cohesion and size of sediments. According to the Hjulström curve, large particles[7] required higher velocities to entrain and remain in transport. Small, lighter and loose particles[8] are easier to entrain and are among the last to be deposited at low velocities.

---

7 Gravel and coarser sand
8 Silt and finer sand

Light cohesive particles[9] are somewhat hard to entrain but remain in transport for the longest time due to their light nature (Ritter et al., 2002). With low velocities propelling particles forward, gravity becomes the dominant transport driver whereby particles will fall to the bed (Hutton and Syvitski, 2008) with the heaviest grain size falling out of transport first. Therefore, what is left on the bed can be indicative of the transport energy at a given time and a crucial component of sediment core interpretation (Hamilton et al., 2015). With the same volume of water, velocity will decrease when depth and width decrease (Leopold and Maddock, 1953). Due to the relationship between transport energy and grain size type, rivers typically experience a fining of material deposited downstream as the river velocity decreases approaching the wide river mouth and receiving basin of an estuary (Ritter et al., 2002).

Grain size fining in estuaries is often more complex than the typical fining trends observed within a river. In an estuary, according to Hjulström's theory, the coarsest material deposits at the river mouth[10], or sediment source, with a fining of material leading away from the river as velocities slow. However, Hjulström's curve does not account for the type and amount of available material in the system (Ritter et al., 2002). Thus, even under low flow velocities, areas that are sediment starved may experience minimal deposition. Additionally, in an estuary, the river velocity is also interacting against ocean currents and tides that can re-entrain, break down, and transport material along the bed. Thus, due to multiple sources of energy for sediment transport, Darlymple et al. (1992) observed in estuaries, the sediments at the bed tend to move landward in the outer zone, be reworked or converge in the central zone, and move seaward in the inner zone. However, suspended sediment does not necessarily display such a clear net movement direction as the bedload (Dalrymple et al., 1992). Also, assess an area using the Hjulström's curve based on current velocities does not account for deposits on the seabed from historic glacial activity, slope failures, or infrequent high energy tsunami events. For example, on the Fraser delta and estuary, slope failure events have been noted to mobilize over $10^6 m^3$ of silty sand off of the delta (McKenna et al., 1992). Thus, coarser deposits than the current transport energy can transport, often remain on the seafloor. Such coarse material can source from storm events and past deposits from glaciers that can move mass amounts of sediment.

---

9 clay
10 where the fluvial discharge is the highest in comparison to the outer estuary

The Impact of Changes in Estuarine Sediment Inputs on Morphology

After a shift in sediment load to the estuary due to basin-wide change in land use or climate, there is a time delay before these changes will become reflected in the morphology (Brouwer et al., 2018). The estuarine turbidity maxima will shift seaward when sediment load is high and landward when sediment load is low depending on the season. This, in turn, will impact deposits to the seabed. Channel infilling can occur within decades when the sediment load increase is high enough (Brouwer et al., 2018). Over multiple years, a decrease in sediment load along with rising sea level decreases intertidal tidal mudflat and coastal wetland areas of the estuary (Maan et al., 2018). Discharge, along with sediment load, plays a role in determining erosion and deposition (Brouwer et al., 2018).

Plume Dispersion

Described in Ritter et al. (2002), the degree of water mixing and type of plume exiting from the river will depend on water density. Salinity, temperature, and sediment concentration impact density. Saltwater is typically denser than freshwater allowing for a salt wedge, saltwater flowing below the river water, and the river plume spreading laterally in a plane-jet[11] flow in the first few depths of the water column. This kind of plume, when saltwater is denser than the incoming freshwater, is referred to as a hypopycnal plume and the mixing of the two water bodies is relatively slow (Ritter et al., 2002). When incoming river water is cold and very turbid, the river water may be denser than the ocean water forming a hyperpycnal plume with plane-jet type flow along the basin floor (Ritter et al., 2002). A hyperpycnal plume is a particular kind of turbidity current[12] that can be locally erosive due to their fast flow can form along the basin floor. If incoming river water and oceanic saltwater are nearly equal in density, homopycnal plumes can form with axel-jet[13] flow (Ritter et al., 2002). In a homopycnal plume, high (near-complete) mixing of the water bodies occur with high deposition rates at the river mouth where the water bodies meet (Ritter et al., 2002). In this way, the density of the water bodies plays a role in determining the spread, deposition, or erosional nature of the river plume.

---

11 Two-dimensional mixing

12 Turbidity currents are rapidly flowing, dense, and highly turbid currents that typically flow along the seabed down a slope (Hage et al., 2018).

13 Three-dimensional mixing

Plume type and mixing can vary between tides. The Upper[14] St. Lawrence estuary is a tidal, flood dominated[15] system that is stratified to moderately mixed (Hamblin et al., 1988). In the Upper St. Lawrence estuary, surface suspended sediment concentrations were typically at a maximum during and just after the ebb cycle (Hamblin et al., 1988). According to Hamblin et al. (1988), upward diffusion of sediments from lower layers was observed with reduced vertical stratification in the water column during the ebb tide when the tidal wedge[16] was less prominent. Near bottom sediments were observed in highest concentrations after the highest currents. The Upper St. Lawrence estuary tidal and riverine interaction shows an environment where contaminants and sediments are recycled throughout the estuary by tidal pumping. The Upper St. Lawrence estuary also shows an example where tidal processes interact with local circulation and water density patterns to create turbulent mixing during the ebb tide (Hamblin et al., 1988). Typically, in macrotidal estuaries, the highest sediment concentrations are observed when there is the most turbulent mixing such as in between the ebb and flood tidal cycle when the ocean and river waters meet with the most strength. The St. Lawrence estuary has a strong tidal wedge causing the time of highest turbulence to occur during the height of the ebb tide rather than between tides when the most salt and freshwater meet (Hamblin et al., 1988).

Turbidity Currents

Hizett et al. (2017) have highlighted how turbidity currents trigger more commonly through hyperpycnal plumes (e.g.: at fjord heads off ice fronts) and submarine landslides off of steep accumulating slopes, but also through sediment settling in hypopycnal flows mobilized by internal waves and tides. Mulder and Syvitski (1995) have found, through an analysis of sediment concentrations and discharge in rivers globally, that turbidity currents have a greater return interval in dirtier rivers. For example, it may take a high flood event in a moderately river to trigger a turbidity current, whereas it may occur every few years in a strong freshet within a dirty river (Mulder and Syvitski, 1995). A study of 95 turbidity current flows on the Squamish delta, found that 73% of the flows were from the settling of river plumes and the remaining 27%

---

14 From the fresher water area near Quebec to the more saline Sagueney Fjord (Hamblin et al., 1988).
15 The flood tide reaches higher velocities than the ebb.
16 A tidal wedge forms when the water column is highly stratified and salt water that is more dense wedges beneath fresh water (Ritter et al, 2002). A tidal wedge forms only during certain periods of the tidal cycle.

were associated with landslide events (Hizett et al., 2017). Turbidity currents leave layers of coarser grains in the sedimentary record and erosional channel features on the seabed (Hage et al., 2018; Gales et al., 2018). Turbidity currents leave layers of coarser grains in the sedimentary record and erosional channel features on the seabed (Hage et al., 2018; Gales et al., 2018). Below the turbidity currents, submarine fans can form where the mass of sediment eroded in the erosional channel is deposited and accumulates (Gales et al., 2018).

Estuarine Physiography and Features

Multiple classifications based on the current morphology and history of how the estuary forms exist that typically include, drowned river valleys, glacially formed, tectonically produced, and bar-built estuaries (Pritchard (1967). However, the primary morphogenetic classification of estuaries presented by Perillo (1995) focus on the drowned river, glacial, and river dominated valleys that are more relevant to the Skeena estuary (Figure LIT2). In the Perillo (1995) classification, drowned river valleys and rias have a drowned V-shaped topography while former glacial valleys such as fjords and fjards are u-shaped. The major difference between fjords and fjards is that the latter are shallow, have highly irregular inner shores, and have lateral tributaries. Estuaries also typically widen and deepen seaward (Friedman and Sanders, 1978). According to Wolanski & Elliott (2007), most of the estuaries of the present deposit sediment into drowned river valleys that existed above surface roughly 20 000 years ago when the global sea level was 120-130 meters lower than today. However, valleys and fjords can also form from glacial activity that was drowned by rising sea level (Paturej, 2008; Perillo, 1995).

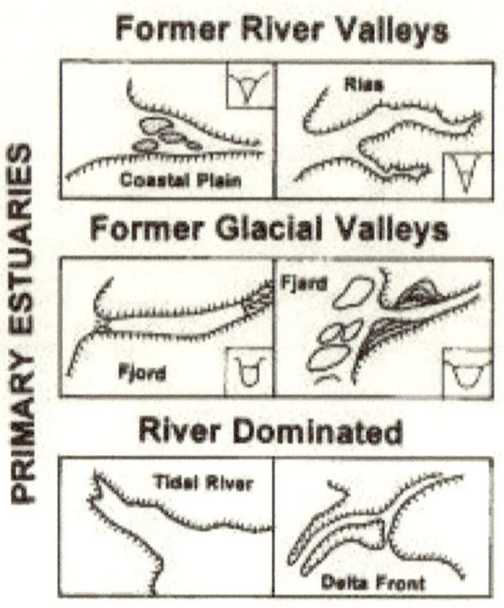

2 Figure LIT2: Morphogenetic classification of estuaries (Perillo, 1995).

According to Dalrymple et al. (1992), regardless of the overall physiography, circulation, or dominant processes, estuaries typically possess a tripartite structure: an outer (seaward) marine dominated portion, a low energy or mixed energy central zone, and an inner (landward) river dominated area (Fig LIT3A). Allen (1991) also describes a tripartite estuarine structure with a meandering upper river channel zone characterized by sandy and muddy point bars, an estuarine funnel with sand bars and vegetated islands, and a constricted and tidally scoured tidal channel inlet. The commonly referenced Dalrymple et al.'s (1992) tripartite structure includes sediment convergence where sediment deposits display a fining trend leading towards a central zone. The outer, marine dominated zone is a section usually in the tidal inlet where net transport and fining is towards the head of the estuary. The central basin is the low energy convergence zone where fine sediments lie. The inner estuary is the river dominated and tidally influenced zone where grain sizes fine seaward from the river. Dalrymple notes that this tripartite structure may vary depending on whether the overall estuarine system is wave or tide dominated and based on the underlying physiography. Furthermore, the tripartite structure is more apparent at the seabed in wave dominated estuaries with lower tidal inputs (Darlymple et al., 1992). In tide dominate estuaries, the tidal energy extends further landward producing a more gradual transition between zones with mudflats lining the banks while coarser sediments and sand bars are present in the channel center (Darlymple et al. (1992).

Tidally Dominated Drowned River Valley Estuaries

Darlymple et al. (1992) model was written primarily based on the Bay of Fundy, a macrotidal estuary with low fluvial inputs and situated within a tidally drowned river valley. Thus, general structures (such as the tripartite structure) may be visible in all estuaries to some degree. However, specific morphological observations described by Darlymple et al. (1992) may not match fjords or other non-drowned river valley areas. For example, shown in Figure LIT3B, in tidally dominated drowned river valleys, estuaries, river meanders before entering the receiving basin are taken as an indication of estuarine conditions due to the mixing of tidal landward and fluvial seaward energy slowing river movement to cause meanders (Dalrymple et al., 1992). Meanwhile, a straight channel entry into the receiving basin is indicative of a delta system that is fluvial dominated (Dalrymple et al., 1992). Once the tidal river reaches the receiving basin, broad parallel-laminated sand flats with braided channel patterns referred to as the upper flow regime sand flats form. Further seaward, elongated sand bars can form that consist of cross-bedded medium to coarse sand. Along the estuary sides in energy minimum zones, muddy sediments accumulate in tidal flats and marshes (Dalrymple et al., 1992). However, Darlymple et al. (1992) highlight that the estuarine zones will shift based on sediment availability, coastal zones gradients, and the stage of estuary evolution.

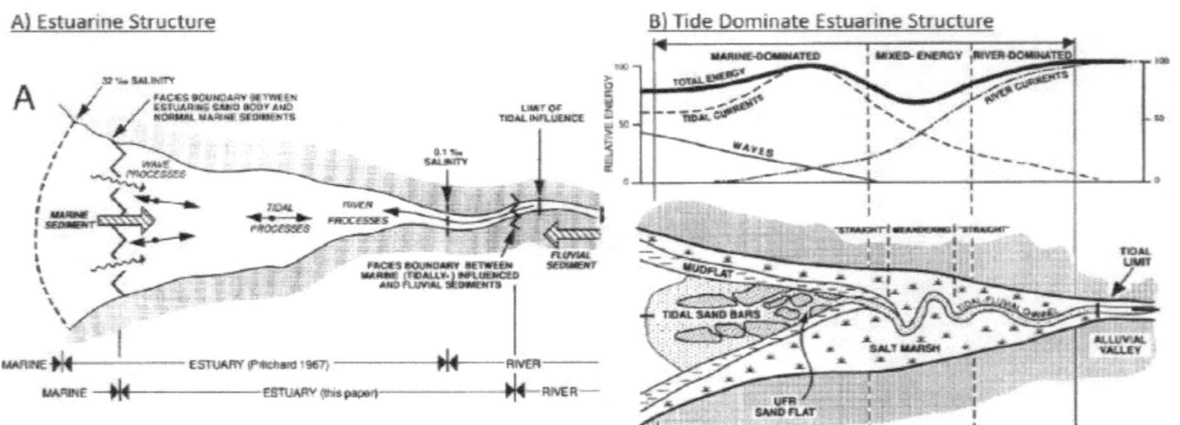

3 Figure LIT3: The tripartite structure of an unconfined and tide dominated estuary modified from Dalrymple et al. (1992). Figure LIT3A shows the general structure of an unconfined estuary with different definitions of estuarine extent based on salinity (Prichard, 1967) and fluvial vs marine sediment facies (Dalrymple et al., 1992). Within the estuary defined zone of Figure LIT3A, Dalrymple et al. (1992) separated marine and river dominated zones as well as mixed energy (Figure LIT3B). Notice that tidal energy extends further than waves into the river-dominated zone. Figure LIT3B also displays the specific morphology and energy zones for an unconfined, tide dominated drowned river valley estuary.

Fjord Estuaries

Although scoured by glaciers forming a deep basin, fjords can also show diverse morphological features and phenomena on their seabed. Deposits of glacial diamicton and glaciomarine sediments intersperse postglacial sediment forming a varied and at times abruptly transitioning seabed (Shaw et al., 2017). Glass sponge reefs, slope failures and fan deltas are visible on the seafloor and steep banks of the Kitimat fjord system (Shaw et al., 2017). Fan deltas form from debris avalanches or incoming tributaries. A subaqueous fjord head delta formed at the mouth of the Kitimat fjord, but due to the deep nature of the fjord, the delta remained subaqueous (Shaw et al., 2017). Fjord head deltas form when a river enters the deep fjord basin, reduces velocity, and drops its coarse sediments. According to Syvitski et al. (2012), talus cones from the steep wall material as well as incoming fluvial sediment continue to infill modern fjords after glaciers retreat. Infilling fjords may have layers of lacustrine mud, fluvial sand and gravel, or marine silt deposited over glacial till from the last glaciation. Glacial moraines and hanging valleys with waterfalls may also be visible within fjord geomorphology. Fjords may also experience seasonal pulses from inflowing rivers allowing for seasonal sediment deposition to the seafloor. Highly turbid and cold freshwater entering a deep, unstable fjord bank can produce plunging hyperpycnal turbidity currents under select conditions (Syvitski et al, 2012). As described previously, due to their deep nature, fjords generally exhibit fjordic two-layer stratification and circulation with freshwater outflowing at the surface and seawater at depth (Hansen and Rattray, 1966).

Bedrock Estuaries

Bedrock islands add variation to the system that can result in deviation from the common descriptions and predictions on how a tidal or wave dominated estuary behave. For example, high waves input to the system from the ocean and outer estuary in winter could locally produce features typical of a mixed or even low wave environment within the shadow of the bedrock island (Wolanski and Elliott, 2007). The convergence of opposing bedrock shores in a narrow channel can cause tidal energy to increase per unit width (Flemming, 2011). Meanwhile, the irregular bed and banks approaching bedrock islands can also increase friction, slow tidal energy, and protect select areas from wave action (Flemming, 2011). In a rock confined estuary in Korea with low fluvial input, Lee et al. (2013) found that mud was transported disproportionally in both direction and magnitude between two of the larger estuary channels, which caused different morphodynamic characteristics within the same estuary. In a system with bedrock islands, the immobile islands are inaccessible as a sediment source. This constrains the system into set bedrock channel pathways. Thus, a bedrock island estuary may deviate from the traditional structure used to describe, classify, model, and understand unconfined estuaries.

Deltas

Dalrymple et al. (1992) define a delta as an area where the river forms a net deposition and progradation of sediments seaward in the receiving basin. Deltas can form both within and outside of estuaries. Estuaries do not have the same consistent net movement of bed sediments seaward found within a delta because fluvial inputs dissipate to the ocean or remain with the estuary, recirculate, and eventually deposit throughout the area gradually infilling the drowned valley (Dalrymple et al., 1992; Boyd et al., 1992). According to Friedman and Sanders (1978), deltas generally form at the mouths of rivers with suspended sediment concentrations of the incoming river exceeding 225 mgL^{-1}. Meanwhile, river inflow concentrations less than 160 mgL^{-1} are typical in estuaries. Areas between these concentrations are considered transitional. However, there are many exceptions.

Shown in Figure LIT4, deltas are also generally considered, tide, wave, or fluvial dominated (Galloway, 1975; Boyd et al., 1992; Ritter, 1978). Delta areas that are wave and tide dominated are often highly destructive deltas where sediment redistributes along the shore in plains and flats marginal to the delta (Ritter, 2002). Shown in the Copper Delta in Figure LIT4, wave dominated deltas typically have sediment protruding from the coast (parallel with the shoreline) with beach ridges more common (Seybold et al., 2007; Galloway, 1975). Shown in the Fly and Mahakam delta in Figure LIT4, tide dominated deltas can look similar to an estuarine bay except with many islands that form parallel to the dominant tidal flows and perpendicular to the shore (Seybold et al., 2007; Galloway, 1975). Conversely, high constructive deltas refer to areas where fluvial action is prevalent and accumulating mass amounts of sediment (Ritter, 1978).

4 *Figure LIT4: Common delta type classifications by Galloway (1975).*

Based on shape, the classic delta formation is triangular, giving the formation its name (Friedman and Sanders, 1978; Ritter, 1978; De Blij et al., 2009). However, shown in Figure LIT4, deltas formations have also been described as lobate[17] or elongate[18] (Ritter, 1978; Galloway, 1975). Some sources separate lobate deltas into fan[19] and braided[20] delta categories (Liangqing and Galloway, 1991). Other sources have described deltas as cuspate[21], birdsfooot[22] (Figure LIT4 Mississippi), or arcuate[23] in their shape (Trenhaile, 2010). Deltas can also classify as dominantly coarse or fine grain formations (Liangqing and Galloway, 1991; Orton and Reading, 1993). For example, Gilbert-type deltas are dominantly coarse grain type deltas that often form in lakes and demonstrate a fining trend leading away from the river mouth (Friedman and Sanders, 1978). Coarse gravel deltas typically have the steepest gradients (Orton and Reading, 1993). Large unconfined deltas, especially sandy and braided delta plains, promote bifurcation of the incoming channel (Orton and Reading, 1993). Mixed load delta plains with

---

17 with a curved and often fan like shape.
18 with a protruding and sometimes birds-foot shape.
19 Sediment deposit in a fan shape often with minimal braiding channels
20 Many braiding channels depositing enough sediment to prograde as a delta.
21 with a V-shape.
22 Similar to a bird's foot. See the Mississippi delta as an example (Friedman and Sanders,1978; Ritter, 1978).
23 with an arc or fan-like shape. See the Niger River delta as an example (De Blij et al., 2009).

meandering channels have low gradients; however, fine-grained delta plains typically have very low gradients (Orton and Reading, 1993). The cohesive nature of clay commonly serves to limit bifurcation by maintaining channel structure (Orton and Reading, 1993).

Regardless of the type and shape of the delta, certain general formations, phenomena, and terminology remain somewhat consistent. Deltas are commonly described first with a delta plain or topset bed that aggrades sediment (Friedman and Sanders, 1978; Ritter, 1978). This is followed by a delta front or foreset bed that progrades outwards and deposits sediments along a slope at the angle of repose. Finally, the prodelta or bottomset bed deposits aggrade below the delta front (Friedman and Sanders, 1978; Ritter, 1978). Typically, the sediments on the delta plain are coarser than those within the prodelta (Friedman and Sanders, 1978). Countercurrents such as those from tides can produce topset deposits. As the delta progrades, coarser foreset and topset deposit on the muddy, finer, bottomset seabed. Compaction occurs over time, and when coarse sediments deposit on top of fine sediments, they can sink into the softer sediment (Friedman and Sanders, 1978). Deltas also often form distributary channels separating various lobes and shift between depositing sediment within different lobes. Channel avulsion can occur where individual lobes are left abandoned to be transgressed by waves and tides (Friedman and Sanders, 1978).

As mentioned previously, the entire formation can be transgressed and redistributed by winds and tides. An increase in wave or tidal action can dissipate large quantities of sediment inhibiting delta formation and transport more material to the areas surrounding the delta (Friedman and Sanders, 1978). In the areas marginal to the delta, the delta marginal plain, waves and tides can redistribute sediments along the shore forming beach plains or mudflats. Alongshore drift leading away from a delta can also increase spit formation. Alternatively, sediment accumulation and delta formation increase when waves are suppressed. For example, with a low mean incoming suspended sediment concentration from the Mackenzie River of only 34.1 mgL^{-1}, the Mackenzie delta formed into the Arctic Ocean. Floating sea ice suppressed wave action enough for the sediment to accumulate and prograde forming a delta (Friedman and Sanders, 1978).

Deposition on deltas occurs due to the reduction in velocity and turbulence as the river meets a standing or tidal body of water (Ritter et al., 2002). The likelihood of a slope failure event increases as the sediment builds on the delta front, and the slope increases (Friedman and Sanders, 1978). Delta formation is particularly sensitive to the gradient between the riverbed and receiving basin, water density, as well as the total amount of riverine sediment input to the receiving basin (Friedman and Sanders, 1978). The steeper the gradient between the river mouth and the receiving basin, the higher the chance of slope failures as sediments accumulate (Friedman and Sanders, 1978). Therefore, slope stability can be a concern when assessing delta features.

Tidal Flats

Local morphology such as mudflats, sand bars, and barrier dunes, can buffer the impact of waves and storms within an estuary and trap sediment. This can create a protected, lower energy environment that is ideal for sediment deposition. Wolanski & Elliott (2007) describe how mudflats can form when turbid estuarine waters inundate at high tide, deposit material, and the material is not removed during the ebb tide. This material eventually accumulates into a mudflat. Macro and meso-tidal estuaries, estuaries that have a large to moderate change in water level between ebb and flood tides, can form mudflats with clear drainage patterns where the strongest ebb tide currents run. Diatoms can play a role in binding the sediment and thereby decreasing the erosion rate whereas as bioturbation, burrowing animals, can loosen sediment and produce the opposite effect. Vegetation can also colonize intertidal areas and produce tidal wetlands or saltmarshes that are submerged or partially submerged at high tide. Vegetation on mudflats can serve to trap fine sediment and enhance sediment settling. Sediment availability, mineralogy, freshwater inflow, tidal variation, wave activity, and storm frequency also affect the formation and continuation of mudflats (Wolanski & Elliott, 2007).

Anthropogenic Impact on Estuaries

Understanding the flux of river sediments and nutrients to the ocean has been set as a goal of the International Geosphere-Biosphere Programme, Land Ocean Interaction in the Coastal Zone (Syvitski et al., 2005). However, sediment inputs to estuaries are changing due to varied watershed wide anthropogenic land-use changes such as damming, mining, deforestation, agriculture, and development (Wolanski & Elliott, 2007). Deforestation can increase sediment flux to estuaries as the discharge is more rapidly increasing the carrying capacity of rivers and reducing the amount of sediment held in plant roots. Changes in climate due to ongoing climate change alter sediment flux from historical rates. For example, increased forest fires due to changing climates cause a lag of increased terrestrial material and ash within watersheds and subsequent estuaries (Kilham et al., 2012). Pasternack and Bush (1998) have observed through an analysis of multiple sediment cores from ten different tributaries that watersheds with 40-50% of land under cultivation showed a doubling of sedimentation in comparison to the pre-settlement deposition rate. In contrast, the Huang He (Yellow) River, the second largest river in the world in terms of sediment load representing 6% of the global river sediment flux to the ocean, has shown a step-down decrease in sediment load from 1950-2005 (Wang et al., 2007). Wang et al. (2007) attribute this decrease in sediment load due to the construction of dams, limiting historical flooding, sediment conservation practices, increased water consumption, and changes in precipitation due to climate change. With decreases in fluvial discharge, sediment, and nutrient inputs, aquatic ecosystems in the coastal Bohai Sea will expect significant morphological, geological, ecological, and biogeochemical changes (Wang et al., 2007). The changes in sediment load have already altered the accretion and erosion rates within the Yellow River Estuary (Cui and Li, 2010). Often multiple factors across the watershed compound to produce changes in inputs to estuaries. These changes then affect estuarine ecosystems and global carbon cycles.

In a study of the global terrestrial sediment flux, Syvitski et al. (2005) have estimated that 12.6 $BTyr^{-1}$ of sediment is brought from rivers to the oceans in modern day with approximately 100 BT of sediment being trapped behind reservoirs. Comparatively, studies on terrestrial sediment flux during the period determined as pre-Anthropocene was estimated at 15.5 $BTyr^{-1}$. (Syvitski et al., 2005). Based on the study by Syvitski et al. (2005) globally, there has been a net decrease in the transport of sediment to the ocean. However, local areas may experience variable impacts on their sediment load depending on the climate and type of development within the area.

Lotze et al. (2006) have highlighted how estuarine and coastal transformations have accelerated over the past 150 to 300 years. In an estuary, turbid river discharge that is high in nutrients and sediment meets with clearer ocean water. This coupling of high light penetration and nutrients creates an optimal habitat for multiple species such as eelgrass and phytoplankton that form the base of the aquatic food chain (Carr-Harris et al., 2015). Therefore, increases or decreases in sediment can have impacts on estuary ecosystem health. Through an analysis of twelve temperate estuaries and coastal seas in Australia, North America, and Europe, Lotze et al. (2006) has also state that human impacts have also increased species invasion, water degradation, and destroyed greater than 65% of seagrass and wetlands. This has resulted in a depletion of >90% of historically economic, structural, or functional species whereby depletion is defined as a 50-90% decrease in relative abundance in comparison to reconstructed historical baselines (Lotze et al., 2006). In turn, water degradation, species invasion, and removal of seagrass can all affect sediment transport dynamics to create feedback loops for ecosystem shifts.

Skeena Background

Geographical Extent of the Skeena Estuary

The Skeena Estuary is a macrotidal estuary with a substantial fluvial influence and heavy wave action during winter storms all situated in a region with a history of tsunami events, landslides, as well as glaciation (Gottesfeld and Rabnett, 2008). Adding to the complexity of the system are several bedrock islands that confine the river and subsequent estuarine channel flow pathways. Shown in Figure SK1, the Skeena River and Estuary is located in northwestern British Columbia near Prince Rupert. The Skeena River running from the east, the Pacific Ocean to the west, the Nass River draining into Chatham Sound to the north, and the Kitimat fjord system connected via Grenville Channel to the southeast surround the Skeena Estuary (Figure SK1). Gottesfeld and Rabnett (2008) state that the tidal influence extends approximately fifty-five kilometers up the Skeena River up to the Kasiks River tributary (Figure SK1). Meanwhile, riverine non-tidal surface currents and changes in salinity have been observed in Chatham Sound as far as Dixon Entrance and southern Clarence Strait (Gottesfeld and Rabnett, 2008).

The immediate estuary discussed in the furthest detail within this research is just downstream of the tidal limit in the river to the portion of the estuary where the Skeena River divides into Grenville Channel, Ogden Channel, and extends into the southern portion of Chatham Sound (Figure SK1). Natural Resources Canada (NRCan), Canadian Hydrographic Service (CHS), the Department of Fisheries and Oceans (DFO), and the Prince Rupert Port authority all water sample, collect sediment cores, and multibeam survey within the deeper portions of the estuary within Chatham Sound. However, the nearshore has been comparatively lacking in data and understudied, making it the priority within this research.

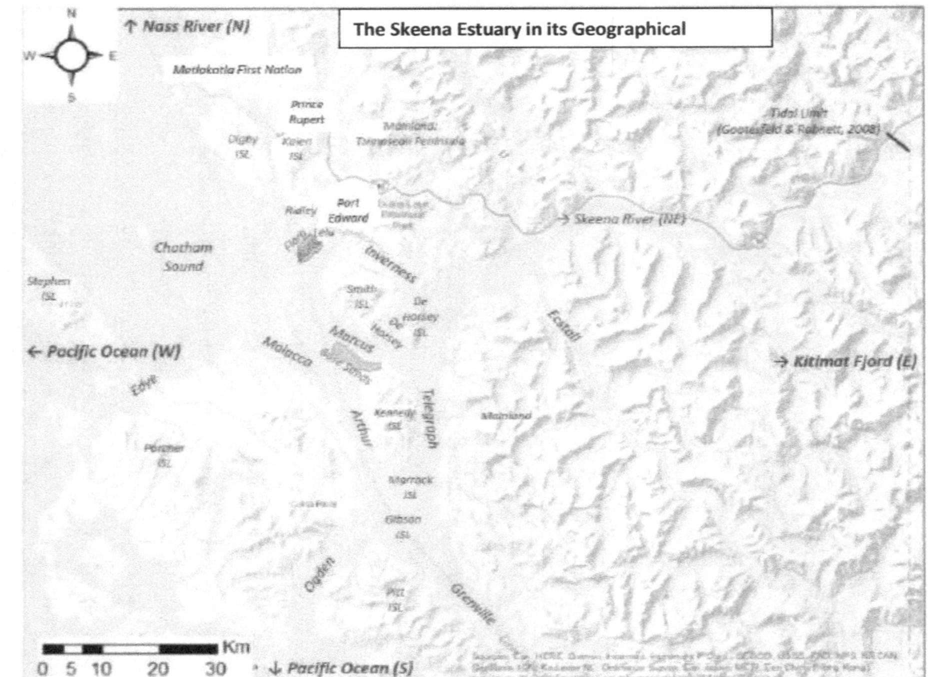

5 Figure SK1: Overview of the Skeena river and estuarine areas with place names.
The maximum tidal influence described by Gottesfeld and Rabnett (2008) is displayed as a red line. Passages, channels, select tributaries, and ocean areas within the estuary are labelled in blue. Islands and mainland areas are labelled in green. Specific banks and bars with a name are labelled in orange. Finally, settlements (towns, cities, and communities) are labelled in black. ESRI layers provided the topographic map base.

Skeena River Hydrology

The Skeena River empties into the Pacific Ocean near Prince Rupert after flowing south and southwest through the Interior (Skeena), Hazelton, and Coast Mountains (Walters *et al.*, 2008; Gottesfeld and Rabnett, 2008). The Skeena River has a watershed area covering ~54,432 km^2 (Gottesfeld and Rabnett, 2008) and a length of 570 km. Climate is considerably different within the coastal (subbasin 1 (SB1) in Figure SK2) portion of the watershed that receives higher precipitation and more moderate temperatures than the interior (subbasin 2 (SB2) in Figure SK2) (Environment and Climate Change Canada (ECCC), 2020). For example, annual precipitation and mean January temperature along the Coast at Prince Rupert is 2500 to 3500 mm and 1.8°C, respectively while in the interior Bulkley Valley at Smithers annual precipitation is 400-500 mm and January temperature is -10.5°C (Clague, 1984). The Coast mountains also produce a rain shadow effect to the east and cause higher precipitation to fall on the western side of the mountains facing the ocean. The furthest downstream hydrometric station directly on the Skeena River is at Usk at river km 450 (Figure SK2). The Usk station excludes a downstream area of 12,050 km^2 (Figure 2 SB1). More specifically, Usk is over ~100 km upstream of the Gottesfeld and Rabnett (2008) tidal limit. Many of the tributaries downstream of Usk that are excluded from the hydrometric station are within the Coast Mountains where substantial rainfall and access to retreating glaciers that could contribute to high discharges and sediment load inputs.

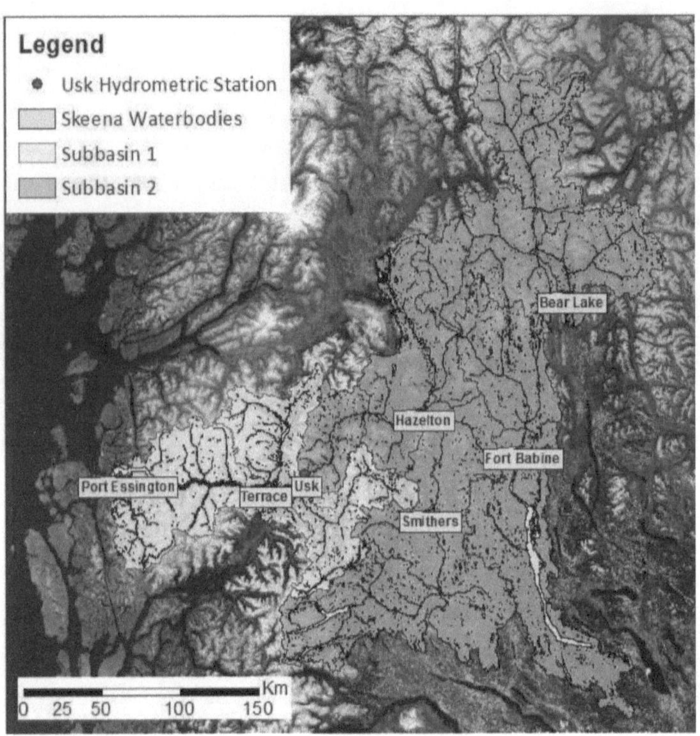

6 Figure SK2: The Skeena watershed with areas upstream and downstream of Usk hydrometric station highlighted in different colors and presented as different subbasins.
The above map was created in ArcMap using ESRI Imagery (2018) and waterbodies & boundaries modified from NRCAN's National Hydro Network (2016).

Discharge is not gauged directly on the Skeena River downstream of Usk; however, gauges are on select downstream tributaries. For the period of 1960-1991, the British Columbia Ministry of Environment (2020) has added the discharge at Usk to measurements of discharge from tributaries with available gauges in the lower reach of the river, including the Zymoetz, Kitsumkalum, Zymagotitz, and Exchamsiks Rivers. This amounted to an annual mean discharge of 2,157 $m^3 s^{-1}$ for the Skeena River (BC Ministry of Environment, 2020). However, the BC Ministry of Environment for Skeena River discharge still excludes all ungauged tributaries including the Lakelse, Gitnadoix, Ecstall, Shames, Exstew, Kasiks, Khtada, Scotia, Khyex, and McNeil Rivers as well as a multitude of other streams and creeks that drain into the Skeena River downstream of the Usk Hydrometric station. The BC Ministry of Environment (2020) mean annual Skeena River discharge appears to be the most complete estimate of discharge for the Skeena watershed available. Other estimates of total discharge for the Skeena are 1760 m^{2}s^{-1} (Benke and Cushing, 2009). In comparison, for the 1981-2010 period, the annual mean discharge at Usk is only 904 m^3s^{-1} (ECCC, 2020).

Binda et al. (1986) described in Milliman and Syvitski (1992), have estimated a sediment flux of 11 million *tyr^{-1}* for the Skeena River with a yield of 260 *tkm^{-2}yr^{-1}* for the 42000 *km^2* (SB2 on Figure SK2) watershed area up to Usk Hydrometric station. Due to the exclusion of SB1, the final sediment inputs from the Skeena River to the Skeena Estuary may be even higher. McLaren (2016) roughly estimated the total sediment load discharged from the Skeena to be 2-5 million *tyr^{-1}*. However, no reference area or methods were described for this estimate. Sediment concentrations within the estuary are highly tidally influenced, altering the water volume per unit of sediment, and making it challenging to discern riverine from marine sediment inputs. Thirteen instantaneous sediment concentrations are available at Usk hydrometric station over various seasons from 1988-1992, resulting in a mean concentration of 00996 *kgm^{-3}*(Environment and Climate Change Canada (ECCC), retrieved 2020). Further, consistently sampling over multiple seasons within SB1 and before the tidal influence on the Skeena River, would be ideal for confirming the sediment concentration of the river and flux to the estuary. However, access is limited, and data is scarce downstream of Terrace and prior to the estuary.

Shown in Figure SK3 and based on Usk Hydrometric station (SB2 on Figure SK2), the Skeena River experiences a spring freshet between May to July (ECCC, 2020) due to snowmelt in the highlands. The mean annual flow during the freshet is 3,160 m^3s^{-1} with an 87 year maximum (1928-2018) of 9340 m^3s^{-1} in 1948 (ECCC, 2020). According to de Groot (2005), the 1948 flood is a 1 in 200-year event. However, this is an estimate based on the available data that is missing a few early flood events and has been measured consistently for less than 87 years. A secondary fall peak in discharge occurs due to high fall precipitation and early winter storms in the coastal portions of the watershed ($< 1000 \ m^3s^{-1}$). The Skeena fall discharge peak is less pronounced than the Nass River as the Skeena river flows deeper into the Interior while the Nass remains relatively coastal for a longer distance (De Groot, 2005). However, the coastal SB1 area would likely add to the fall discharge, when included with SB2 on a hydrograph.

Glacial melt contributes to higher flows in late summer, and lowest flows occur between January to March (150 m^3s^{-1}). In some years, the fall discharge is more pronounced than others, and some years, the freshet may be earlier or later than the mean (ECCC, 2020). Thus, the mean trend averages out the variations in the timing of the freshet producing a low mean freshet (~3000 m^3s^{-1}) over a longer May to July period. With a 2 year return interval, the maximum instantaneous discharge at Usk regularly reaches above 4000 m^3s^{-1} (De Groot, 2005). Based on the Usk Hydrograph over ~87 years of data, flows of 100-1,000 m^3s^{-1} can generally be considered lower flows (e.g. winter flows), 1,000-3,000 m^3s^{-1} moderate (e.g. fall discharge peak), over 3,000 m^3s^{-1} are considered higher (e.g. freshet) flows. This general classification of river flow will be referred to within this book for describing the riverine conditions at Usk over survey dates in the estuary. Over 7000 m^3s^{-1} discharge at Usk historically reported as flood conditions (de Groot, 2005; ECCC, 2020). However, all these values represent the SB2 watershed area, and therefore, flood events, including SB1, are likely even larger.

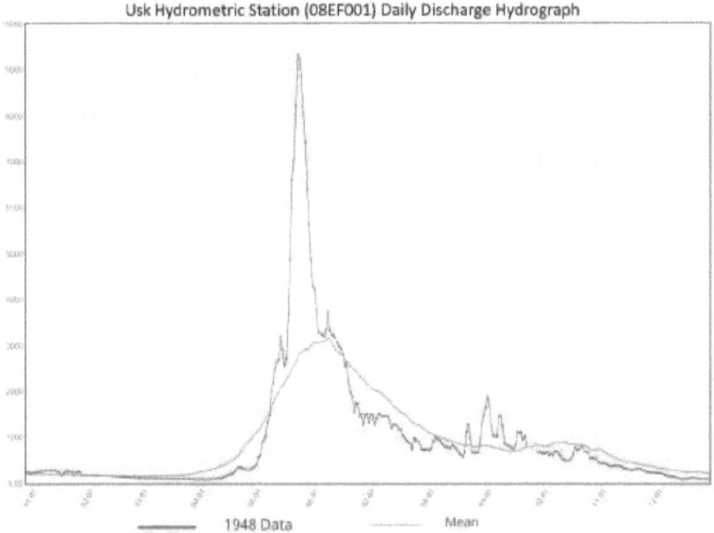

7 Figure SK3: Usk hydrograph modified from Environment & Climate Change Canada (ECCC). The red line displays the daily discharge in the 1948 flood year to represent extreme (~1/200 yr.) discharge conditions. The golden line represents the daily mean discharge from an 87 year (1928-2018) period.

Skeena Basin Geology

Shown in Figure SK4, the interior and coastal portions of the Skeena River watershed encompass different substrates and bedrock types. Upstream of the Coast Mountains, the Skeena river flood plain is wide (~ 5000 - 20000 m in areas) and consists of alluvium and glacial till overlying sedimentary and volcanic bedrock (Cui et al., 2017). The interior drainage consists of more generous proportions of low-grade metamorphic and sedimentary rocks than the predominantly higher grade granitoid and gneiss complexes of the coastal regions (Hutchinson, 1982; Fulton, 1995; Gottesfeld and Rabnett, 2008). According to Crawford et al. (1987) and Clague (1984), between Prince Rupert and Terrace (the majority of subbasin 1 (Figure SK2)) consists of the Coast Plutonic Complex, that is a part of the Coast-Cascades belt and developed between the mid-Cretaceous and mid-Eocene, whereby high temperature gneiss thrust over older low grade Jurassic sedimentary and volcanic rocks of the Intermontane belt to the east. The higher-grade portion of the Skeena watershed is also considered part of the larger Coast

Mountain batholith that extends up and down the northern Coastline of British Columbia (Gehrels et al., 2009).

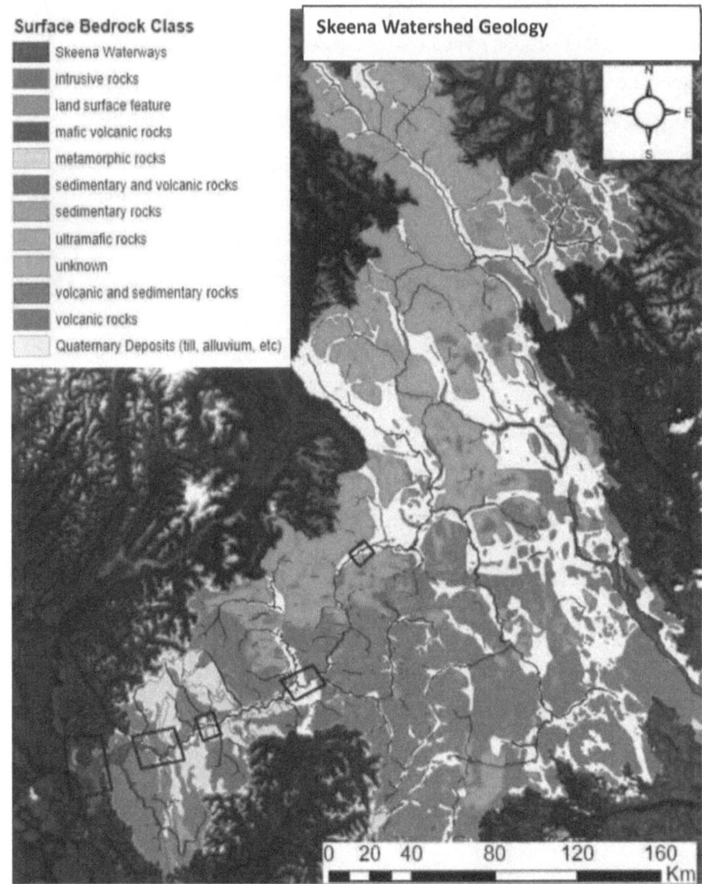

8 *Figure SK4: Surficial Geology of the Skeena watershed.*
This map was created using digital geology layers from Cui et al. (2017). Black boxes refer to different morphological areas of the river described later in this book.

Skeena in the Quaternary Period

Focusing on changes within the Quaternary period, the steep walled fjord of the Skeena River mouth was a major conduit for ice from the Coast Mountains (Clague, 1984). The Skeena watershed area was glaciated during the most recent glaciation (Fraser Glaciation) 28 to 11 ka

(Gottesfeld and Rabnett, 2008) and deglaciation in the region occurred between 13 to 10 ka (Clague, 1984). This glaciation resulted in intensely scoured valleys and the accumulation of tens of meters of glacial till in the lowland areas. The sea inundated the river valley at the end of the last glaciation between 10.9 and 10 ka (Conway et al., 1996). During the deglaciation phase, shorelines were higher than at present causing gravel to accumulate in the Skeena valley while sea level rose above depressed coastal areas resulting in inundation and marine deposits as far as the Zymoetz River (Gottesfeld and Rabnett, 2008). According to Clague (1984), isostatic rebound during early postglacial times then caused rapid marine regressions reaching present sea level around 8 – 8.5 Ka. In the nearby Kitimat fjord area, shorelines fell rapidly around 10.5 ka from about +200 meters to around present-day levels by 8 ka (~8 $cmyr^{-1}$). Archaeology around Prince Rupert Harbor indicates that 40 First Nations villages remain near the present shoreline or several meters above present highwater for over the past 5000 years. Continuous human occupation around present sea level indicates that over the last 5000 years, sea level has been roughly equal to or below the present sea level. In Bella Bella, sea level was lower during portions of this period. Thus, sea level may have fallen below present levels after 8 ka; however, no studies have been conducted around Prince Rupert to confirm this. If sea level were lower, a small transgression to present sea level would have been necessary during mid-late post glacial times (Clague, 1984).

Shown in Figure SK5, drill hole records (1-7) along the banks of the Skeena River approaching the estuary, display the glacial history of the Skeena region (Clague, 1984). More recent alluvial deposits overlay a history of marine deltaic deposits, as well as marine and glaciomarine mud from when the glacial incised bedrock was inundated and infilled (Clague, 1984). The Skeena River and delta prograded westward after inundation leaving behind the alluvial and deltaic deposits. In a few of the deeper drill holes upstream near Terrace, diamicton is present from the glaciation period. Select drill holes also reach down to bedrock. The thickness of the alluvial deposit generally decreases in size as the cores progress downstream (Clague, 1984).

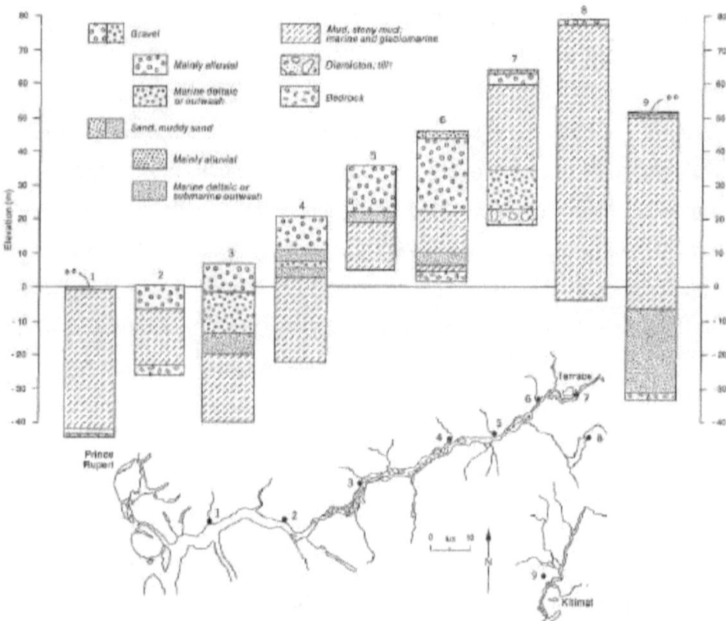

9 *Figure SK5: Clague (1984) drill hole records displaying the late Quaternary stratigraphy for the lower Skeena Valley.*

Today, current vertical land motion in Prince Rupert is considered negligible (British Columbia Ministry of Environment, 2016). However, select areas of the Pacific north coast in southern Alaska, still undergo appreciable isostatic rebound caused by recent ice loss (Larsen et al., 2004). However, uplift may also be masked by diastrophism in the Skeena area whereby evidence has suggested that due to the plate tectonic regime, Vancouver and Haida Gwaii islands are rising while the mainland is stable or subsiding (Clague, 1984). Thus, uplift and subsidence in the Prince Rupert area may be cancelling out one another, thereby producing the negligible vertical land movement in modern times.

Based on records from 1901-2014 for the tidal gauge at Prince Rupert, the average relative sea level rose at 13.3 cm per century ($\sim$1.33 $mmyr^{-1}$) (British Columbia Ministry of Environment, 2016). Under the moderate representative concentration pathway (RCP) 4.5 estimations, projected sea level relative to 1995 increases to 3.2 $mmyr^{-1}$ by 2040, 3.88 $mmyr^{-1}$ by 2070, and 4.419 $mmyr^{-1}$ by 2100 for the Prince Rupert area (James et al., 2015).

This results in a total median sea level rise of 46.4 cm by the year 2100 under RCP 4.5. Under the more intense 8.5 RCP estimations, projected relative sea level rise increases at 3 $mmyr^{-1}$ by 2040, 4.35 $mmyr^{-1}$ by 2070, and 5.495 $mmyr^{-1}$ by 2100 (James et al., 2015). Thus, over the past century the Skeena area has experienced sea level rise and sea level rise is expected to increase through the 21st century.

Skeena Estuary Hydrodynamics

The immediate estuary is macrotidal with large tidal fluctuations and intensive mixing. Tides are semi-diurnal (Trites, 1952). Tides originate from the amphidromic point in Hawaii (Thompson, 1981), approach the area from the south, and propagate counterclockwise from the northwesterly trending coastline. At the Prince Rupert long-term station, the highest tide is at +7.98 and the lowest at -0.38 m (Fisheries and Oceans Canada, 2013). This produces a maximum range of 8.4 meters. The mean level from 1939-2020 is at 3.86 m above chart datum (relative to mean low tide in Canada) (Fisheries and Oceans Canada, 2015).

Limited data are available regarding Skeena Estuary wave dynamics. The ShoreZone project has mapped the wave exposure of the coastline within the Skeena Estuary (Figure SK6). According to the ShoreZone map (1998-2000), the inner channels approaching the river are protected while Chatham Sound is semi-protected to semi-exposed; although, some of the wave effects amplify as the seabed decreases in depth approaching the delta. According to Hoos (1975), although waves may be locally produced, the inner delta and tidally drowned river are generally buffered from waves and waves are limited in fetch by the bedrock islands except during larger winter storms. The primary exposure of the estuary to waves occurs when winds approach Chatham Sound from the northwest and are amplified by the approaching delta slope in outer Marcus Passage (Hoos, 1975). Thus, the Base Sands area (Figure SK1) can experience waves of approximately 0.5 to 1 m, even without a storm when the winds come from the northwest (Hoos, 1975). Telegraph Passage may also experience substantial wave action, but less information is available for the area. However, the most frequent wind direction in Prince Rupert from October to April is from the southeast, from the South in May, August, and September, and from the west in June and July according to the 1981-2010 Canadian Climate Normals (ECCC, 2020). Based on the Prince Rupert dataset winds will be coming more frequently from the interior and pushing seaward out of the estuarine passages. Thus, winter arctic outflow storms would likely have the most significant impact on waves within the estuary.

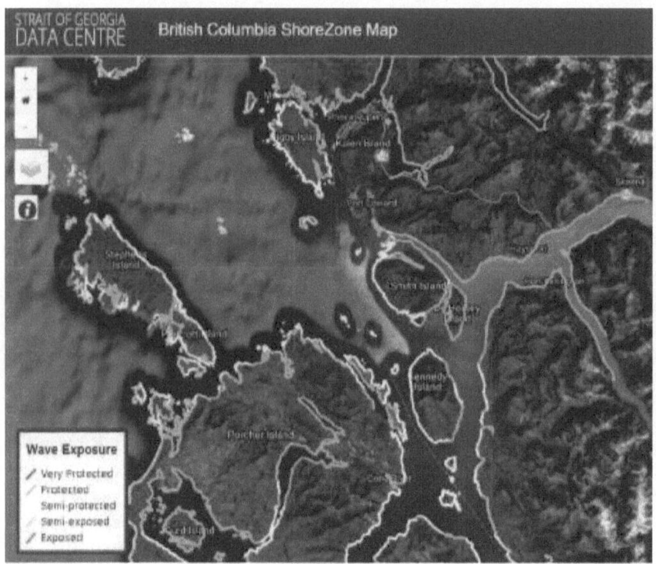

10 *Figure SK6: BC ShoreZone (1998-2000) Wave Exposure in the Skeena Estuary area.*

Extensive studies on salinity by Trites (1956) and Cameroon (1948) have been conducted and compiled by Hoos (1975). The compilation by Hoos (1975) will be predominantly referred to within this report as the original Cameroon (1948) document was inaccessible. Percent freshwater is a term used by Hoos (1975) to describe the samples departure from average marine salinity conditions and does not mention the PSU salinity. Instead, percent freshwater is derived based on the average upper 60 ft (15 m) of the water column salinity and the mean seawater salinity. Thus, an area that is 20% freshwater can be determined through the reverse calculation to be approximately 28 PSU. According to Hoos (1975), based on June and August samples, Inverness, Telegraph, and Marcus Passages as well as the area towards the Skeena River experience freshwater contributions > 20% in the upper 15 m (Fig SK7). Under non-freshet conditions, a more saline zone develops connecting Chatham Sound and Ogden Channel through Malacca and Arthur Passages. During freshet conditions, the equivalent percent freshwater zone pushes to the marine edges of Chatham Sound (Hoos, 1975). The east portion of the estuary is consistently less saline than the marine exposed west. Under moderate to low river inflows percent freshwater downstream of the convergence with the Ecstall tributary is > 90% at the convergence of the Ecstall tributary, >60% landward of the entrance to Inverness, and 20-60%

within Telegraph passage until the southern end of Kennedy Island. In the further reaches of Telegraph Passage at depths greater than 15 m, salinity gradients drop below 5% freshwater (Hoos, 1975).

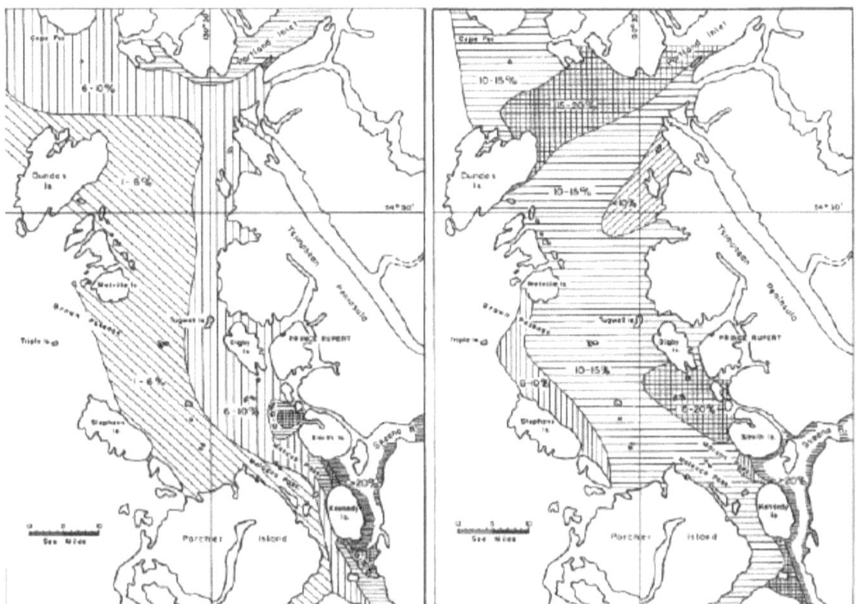

11 *Figure SK7: Hoos (1975) August (left) and June (right) percent freshwater in the upper 15 m of the Skeena Estuary waters.*

Skeena Estuary Seabed Morphology

In terms of morphology, Hoos (1975) describes the Skeena River mouth as deltaic. Conway et al. (1996) also describe the Skeena Estuary as a delta with several depocenters separated by islands and shoals that prograde into deep water. Luternauer (1976) defined the extent of the Skeena River delta front based on a transition point in the percentage of sand opposed to mud on the seafloor (Fig SK8, left). He used seismic profiles and grab samples collected and sieved by Natural Resources Canada in 1974 to delimit the deltas extent. Luternauer (1976) noted a break in slope typical of a delta front with broad sandy flats at the delta edge between Ridley to Gibson Island (topset facies). However, the delta extent by Luternauer includes limited data within Inverness Passage. In 2016, McLaren collected sediment grab samples at 2,467 locations to analyze grain size distribution within the estuary near Inverness Passage and Flora Bank (Fig SK8, right). Within the outer portions of Inverness Passage, McLaren (2016) describes hitting predominantly bedrock rather than deltaic sand. Further seaward, many of the seabed sediments in Flora Bank area appear as reworked glacial material (McLaren, 2016). Flora Bank was categorized as a glacial relic deposit rather than deltaic, due to the unconsolidated nature of the sediments and the 70% similarity of the primary Flora Bank grain size distribution to a till deposit on land (McLaren, 2016). Samples from both McLaren (2016) and Luternauer (1976) were collected from a boat and may underrepresent the shallow areas approaching the banks. Shown in Figure SK9, mud and sand flats, as well as gravel deposits and bedrock, have been visually observed along the estuary banks from aerial photographs taken at low tide between the late 1980s to 2007 by BC ShoreZone (1998-2000). In conclusion, past glacial deposits may also influence on the morphology of the area, and there is still some uncertainty on the extent of the subaqueous delta.

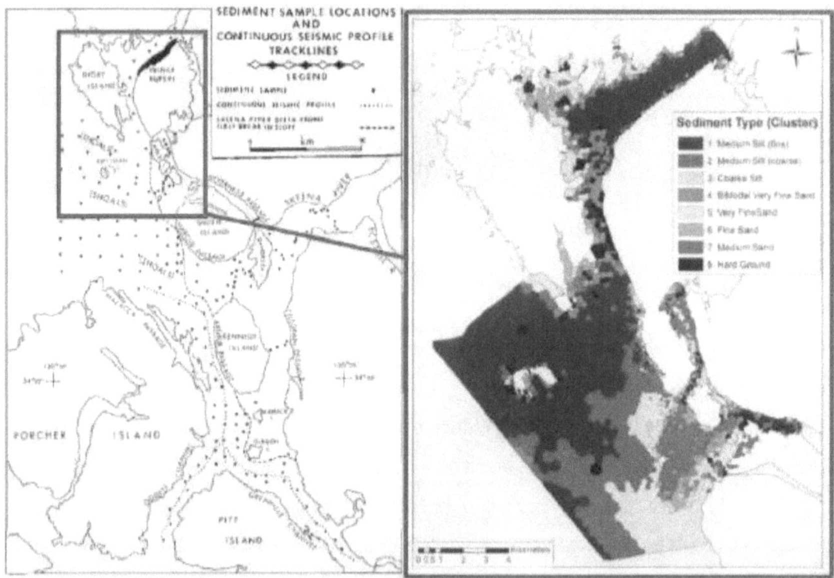

12 *Figure SK8: Skeena Estuary seabed delta extent and grain size deposits from past literature. Right: Skeena River Delta Front and sample locations slightly modified from Luternauer (1976). Left: Grab samples taken of the seabed surrounding the final reach of Inverness Passage and Flora Bank modified from McLaren (2016).*

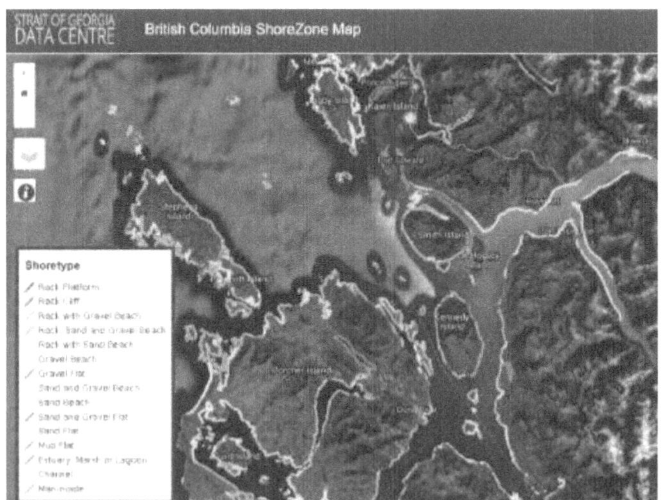

13 *Figure SK9: BC ShoreZone (1998-2000) shoreline based on low-tide aerial photographs.*

On a smaller scale, slope failures, earthquakes, and Hexactinellid sponge reef may also influence the morphology of the seabed in the Skeena Estuary. Debris flows occur regularly along slopes between Prince Rupert and Port Edward during heavy rainfall events (Clague, 1984). Some of these debris' flows have blocked coastal roads or destroyed homes (Clague, 1984). Within the estuary, slope failures have been known to frequent Inverness Passage[24] (Gulf of Georgia Cannery Society, 2020; Wishart, 2018). Nearby Haida Gwaii is an active earthquake zone that is just offshore of the Skeena mainland (Clague, 1984). Several fault lines within the estuary have been documented by the British Columbia Geological Survey (Cui et al., 2017). The region could also be affected by larger offshore earthquakes that may be reflected in the seabed morphology. Hexactinellid sponge reef area described by Shaw et al. (2018) are contributing to the seabed topography and dispersion of sediment in Chatham Sound.

Local Relevance of Sediment Load and Morphodynamics

The Skeena River plays an influential role in the sediment and nutrient conditions of the Skeena Estuary (Fisheries and Oceans Canada, 2015) and can influence eelgrass and juvenile salmon habitats (Lee *et al.*, 2005; Moore *et al.*, 1997). Young salmon cannot travel far in clear

---

[24] According to Wishart (2018), the Tsimshian word Wilałoo for Inverness Passage in the Skeena Estuary means the place of landslides.

waters before starving or being preyed upon (Gregory and Levings, 1998). Three hundred seventy-six million juvenile salmon leave the mouth of the Skeena River and use the semi-turbid estuary to forage and camouflage especially in eelgrass areas that offer food and shelter (Ocean Ecology, 2014). Eelgrass habitats are also sensitive to incoming suspended sediment as eelgrass coverage has been known to decrease when the concentration of fine sediments increases due to reduced light penetration (Lee *et al.*, 2005; Moore *et al.*, 1997). Without a proper understanding of riverine sediment load, it is difficult to explore what conditions could evoke an irrevocable regime shift in habitat and morphology within an estuary. Inaccurate, exclusionary, or data lacking estimates of sediment inputs to the estuary could produce incorrect predictions on what is considered stable and non-threatening to estuarine habitats. In turn, this could lead to false assumptions on contaminant dispersion, sediment infilling, and nutrient cycling.

McLaren (2016) conclusions on the glacial origin of Flora Bank sparked a debate on the deltaic nature and extent of the Skeena Estuary as well as the stability of the bank and subsequent eelgrass habitat. Furthermore, conflicting results within the McLaren (2016) study and corporate Delft3D modelling results surfaced and raised concerns regarding bank stability and eelgrass habitat preservation under proposed[25] LNG terminal developments over the Flora Bank area (Fisheries and Oceans Canada, 2015; Skeena Watershed Conservation Coalition, 2014). Increasing development proposals within the past few decades, the importance of the estuary to wildlife and waterfowl, the significance of the fishery to the livelihood of local communities, the presence of the Port of Prince Rupert[26], the lack of data within the nearshore, and conflicting results within the currently available literature all highlight the local relevance of further studies on Skeena Estuary morphodynamics.

---

25 Flora Bank LNG proposals have since been withdrawn or cancelled according to the British Columbia Ministry of Natural Gas Development (ND) and Northwest Institute & SkeenaWild Conservation Trust (2018).
26 The second largest port with access to Asia in Canada.

The book has been divided into three chapters to assess the impact of bedrock confinement on the estuarine system and to address inconsistencies within the current research available within the Skeena Estuary region. As the current knowledge gap is greatest within the nearshore areas of the estuary, this research focusses primarily on the estuarine nearshore. However, it also incorporates the broader context including portions on both landward and seaward areas of the estuary. Three research question, split into three chapters, are addressed within this research:

1) What are the current riverine inputs to the Skeena Estuary, and how are these inputs expected to change under different climate change scenarios within the next 100 years?

2) How does bedrock confinement influence estuarine classification and morphodynamics at the seabed within the Skeena Estuary?

3) How does bedrock confinement influence the diversion of river flow and suspended sediment as well as stratification within the water column?

Chapter 1: Skeena River Sediment Load and Discharge under Changing Climate

Introduction

Chapter 1 investigates riverine sediment load and discharge inputs to the estuary from the Skeena river mouth under the historic 1981-2010 climate normal and up to the year 2100 under predicted changes in climate. More specifically, the model HydroTrend is used to estimate incoming discharge (Q), suspended sediment concentration (SSC), suspended sediment load (Qs), and bedload (Qb) for the Skeena River including a previously excluded, due to lack of data, area of the watershed. The models are evaluated over the 1981-2010 period using ECCC hydrometric data at Usk and water samples collected at the tidally drowned river mouth prior to simulating 2010-2100 future conditions. The HydroTrend model is used to estimate riverine inputs to the estuary, first under a past Environment Canada climate normal period (1981-2010), and then up to the year 2100 using two global climate models (GCMs) downscaled with the BCCAQv2[27] method under two different representative concentration pathways (RCPs). No hydrometric stations are available for the final ~150 km reach of the river that would capture the full sediment load (S) and discharge (Q) of the river prior to tidal inputs. The final tributary, Ecstall is already highly tidally influenced within its lower reaches before joining the Skeena (Gottesfeld and Rabnett, 2008). Thus, a model is a useful tool to capture the full watershed area and isolate the river's input prior to tidal influence.

The HydroTrend Model

HydroTrend 3.0.2 uses a water balance model and semi-empirical relationships to estimate glacial and non-glacial suspended sediment load, bedload, and water discharge at a river outlet based on inputs provided by the user that describe the watershed's physical properties and climate (Kettner and Syvitski, 2008; CSDMS, 2018). HydroTrend produces results at a daily time scale that can be post-processed further into user specified time intervals (Kettner and Syvitski, 2008). HydroTrend can also provide inputs for further sedimentation models such as SedFlux, a sister model hosted by the same Community Surface Dynamics Modelling Systems (CSDMS) community. The model has been applied on studies across the globe and reproduces a range of magnitudes in sediment load, including peak flood events (Kettner and Syvitski, 2008). HydroTrend has been used to study fluxes of terrestrial material to the sea given different climate

---

[27] Bias Correction with Constructed Analogues and Quantile mapping, Version 2

and basin inputs (Milliman and Syvitski, 1992). Within Canada, HydroTrend has been used to estimate sediment load and discharge of the Mackenzie River and its tributaries within Natural Resources Canada (Lintern and Haaf, 2014). The model also meets the standards of the American National Standards Institute (ANSI) (CSDMS, 2018).

As described by Kettner and Syvitski (2008), total discharge ($\bar{Q}$) is calculated using a water balance model that simulates precipitation, storage, reduction (e.g.: evaporation), and release through components of rain (Q_r), snow melt (Q_n), glacial melt (Q_{ice}), groundwater discharge (Q_{gr}), and reductions from total evaporation (Q_{eva}) applied over the entire basin area and simulation time.

Long-term suspended sediment load ($\overline{Qs_T}$ or referred to as Qs within this work) is calculated through a combination of the sediment contributions based on climate and basin wide properties ($\overline{Qs_{BQART}}$) and from active glaciers within the basin ($\overline{Qs_G}$) (Kettner and Syvitski, 2008). For the Skeena, Qs_{BQART} is calculated using the BQART method for basins with annual temperatures over or equal to 2°C. The BQART method is described as $\overline{\omega}B\bar{Q}^{0.31}A^{0.5}RT$, whereby, ω is a 0.02 $kg\ s^{-1}\ km^{-2°}C^{-1}$ coefficient of proportionality, long term discharge ($\overline{Q}$), maximum relief (R), drainage basin area (A), average basin temperature, and B (Kettner and Syvitski, 2008). B is a product of the lithology factor (L), trapping efficiency of reservoirs (Te), and the anthropogenic factor Eh through the equation B=L(1-Te) Eh (Kettner and Syvitski, 2008). The sediment contribution from active glaciers within the basin, $Q_{S_G} = (1 - \frac{VS_G}{P_g})\frac{\sum_{a=1}^{n}(1.93*10^{-3}A(9.8)^b - Qs_{BQART})}{n}$, is determined from the annual water storage of rain and snow turned into ice (VS_G), total annual precipitation falling on the glacier (P_g), and the glacial derived annual suspended sediment ($1.93*10^{-3}A(9.8)^b - Qs_{BQART}$) over the number of years (n) of the model simulation run (Kettner and Syvitski, 2008). A is drainage basin area, b logarithm of percentage glacial cover of the basin, and $Q_{S_{BQART}}$ suspended sediment derived from the BQART method. The rationale behind the glacial derived annual suspended sediment is that sediment flux of the river increases with the glacial area (Kettner and Syvitski, 2008). Finally, a Psi model is used to calculate daily values and capture the inter and intra annual variability in Qs (Kettner and Syvitski, 2008).

Bedload (Qb) is calculated using a modified Bagnold's equation, $Qb_{[i]} = \left(\frac{\rho_s}{\rho_s - \rho}\right) \frac{\rho g Q_{[i]}^{\beta} s e_b}{g\,tan\lambda}$ $when\ u \geq u_{cr}$, based on delta slope (s), daily mean discharge ($Q_{[i]}$), sediment density (p_s), fluid density (p), bedload efficiency (e_b), dimensionless bedload rating term (β), limiting angle of repose of sediments on the river bed (λ), stream velocity (u), and critical velocity needed to initiate bedload transport (u_{cr}) (Kettner and Syvitski, 2008).

Methods
HydroTrend Setup for the Skeena Watershed over a Historic Period (1981-2010)
Different approaches were necessary for the final suspended load vs bedload estimates for the Skeena River due to the sensitivity of the bedload vs suspended load HydroTrend equations to different variables. HydroTrend produces outputs at the river mouth defined by the user rather than simulating sediment transport, erosion, deposition, and load in each cell of the river as is the case in specific models. Since the HydroTrend model suspended sediment load results at the river mouth are highly sensitive to climate and lithology inputs, in order to best capture the difference in the coastal and interior regions of the watershed, discharge and suspended sediment load were modelled by applying the HydroTrend model over two subbasins (subbasin 1 (SB1) and subbasin 2 (SB2)) divided at Usk hydrometric station (Figure SK2). Subbasins separated at Usk Hydrometric station provide an opportunity to compare SB2 model results against measured Hydrometric station data and capture differences in climate (discussed later) and lithology (Figure SK4) within the coastal and interior portions of the watershed. The output at each subbasin mouth is combined during post processing to form the final output of discharge and suspended load at the river mouth. This will be referred to as the subbasin combination approach and allows for greater inclusion of basin wide climatic and lithographic variation. Except for its effect on daily discharge, HydroTrend calculates bedload without directly using climate inputs, but based on daily discharge, river mouth slope, mean river velocity, and density (see the previous section on the HydroTrend model). Velocities and slope decrease approaching the final receiving basin lowering the total average river velocity and river mouth slope used to calculate bedload. Some material estimated with a higher velocity and slope in an upstream subbasin would likely fall out of transport prior to meeting the final river mouth.

Thus, adding the bedload from each subbasin simulation (with higher slopes and velocities at arbitrarily assigned subbasin river mouths) would likely over-predict the final bedload of the entire basin (Albert Kettner, personal communications, May 2019). Therefore, the bedload estimate from one run of the model over the entire Skeena basin area rather than through multiple subbasins is more applicable. This will be referred to as the one averaged basin approach and allows for one final calculation of basin bedload using average velocity and slope of the actual final river mouth. Furthermore, during both the historic period and into the future, the basin combination approach is used to estimate suspended load and discharge while the one averaged basin approach is used for bedload.

Each run of the HydroTrend model uses over 23 inputs as well as basin hypsometry from a DEM. Inputs have been adjusted and calculated for each subbasin run of the model. Refer to Table 1Ch1Appx in the Ch1 appendix for a thorough description of the model inputs for each subbasin (values and sources used). Essentially, precipitation and temperature, lapse rate, glacier ELA, groundwater storage, base flow, hydraulic conductivity, average velocity, and other basin characteristics (canopy interception, dry precipitation evaporation fraction, and more) are required inputs for the model and were derived from online datasets or directly referred to within literature. The model does incorporate a general anthropogenic factor, but further research would be necessary to assess the impact of land use changes, forest fires, and more on Skeena sediment load and discharge. Within the model, climate variables produce the most considerable impact on model outputs and will be described further under the model validation section. Climate inputs were derived using data from ECCC climate stations (modelling historical discharge and load) or global climate models (for future discharge and loads). ECCC climate station data with over ten years of data within the 1981-2010 period average monthly into spatially averaged basin wide inputs of precipitation and temperature to compare to the GCM climate variables (Figure 1Ch1 and 2Ch1). Eight ECCC Climate stations are available in the Skeena River subbasins with over ten years of consistent data and climate normal averages within the 1981-2010 period. The climate normal data was averaged monthly and annually for each subbasin. The HydroTrend model was then run over a 30-year interval for validation with the ECCC Usk 30 year climate normal of hydrometric data. Although the model produces daily outputs, the data is summarized as 30 year mean annual or monthly values in the results.

Presented within the Ch1 Appendix under the HydroTrend Input Uncertainty Analysis section, variables that had multiple potential inputs underwent an uncertainty analysis whereby HydroTrend Sb2 outputs compare against measured Usk Hydrometric station discharge (Q). The input producing HydroTrend outputs closest to the Usk Hydrometric station values was chosen moving forward with the production of final outputs for the Skeena. Two such variables were glacial change per year (glacier retreat) and dry precipitation evaporation fraction (ICE) for the 1981-2010 climate normal are described and shown in Table 2Ch1Appx within the Ch1 Appendix. Once final inputs were selected, the historical model run using ECCC climate inputs over SB2 was used to validate the model against Usk Hydrometric station to assess model performance for the Skeena watershed and compare to GCM HydroTrend simulations for the same historic 1981-2010 period.

Methods for using Different Climate Scenarios to Model into the Future (2010-2100)

Climate variables and the glacial change parameter were the only inputs changed to model Q, Qs, and Qb under future climate change scenarios. Based on data available, all other inputs, including basin hypsometry, were held constant to the values used in the historical (1981-2010) simulation (Table 1Ch1Appx). Statistically downscaled climate raster 300 arc second grids (0.0833 degrees or ~10 km) for each chosen GCM were downloaded every five years, to increase computational efficiency, clipped over the subbasin, and averaged into 30-year monthly values from the Pacific Climate Impacts Consortium data portal (PCIC, 2019). Refer to Table 1Ch1Appx for further details on sources and data input preparation. Downscaling of the GCMs was performed by PCIC using the BCCAQv2 method (PCIC, 2013). PCIC (2019) downscaled climate data are available for multiple GCMs and RCPs based on Canadian historically gridded climate data and GCM projections from the Coupled Model Intercomparison Project Phase 5 (CIMP5). Climate grids are not constrained to reproduce natural El Nino-Southern Oscillation climate variability in observations (PCIC, 2019). Each GCM and RCP combination required data preparation (averaging into 30-year intervals) and model simulation for each subbasin. Once again 30-year simulations of the model were chosen to match with ECCC climate normal periods. Thus, final simulations of future Skeena sediment load and discharge present the 2011-2040, 2041-2070, and 2071-2100 periods under two GCMs and two RCPs (selection of GCMs and RCPs described further below). This produces 12 river mouth outputs of potential future conditions (4 for each climate normal period) excluding the 1981-2010 historical period validation.

Global Climate Model (GCM) Selection

The top three climate models recommended for Western North America (WNA) (Giorgi and Francisco,2000) by the Pacific Climate Impacts Consortium (PCIC), CNRM_CM5-r1 (referred to simply as CNRM), CANESM2-r1 (often referred to as only CANESM), and ACCESS1-0-r (referred to as ACCESS), were chosen for this study. The 12 GCMS listed for the Skeena region capture 90% of the variance between all GCMs made for the area (PCIC, 2013). The PCIC website list climate models in an order that allows the user to capture the broadest range between climate models for a given region (T. Murdock, personal communications, March 10, 2020). Three GCMs were initially chosen as a minimum requirement to capture climate variability between GCMs. Six models were chosen for a modelling study within southwestern Canada (Werner, 2016) and the top three models are the same as those listed by PCIC and used within this research.

Representative Concentration Pathways (RCP) Selection and Description

Two RCP scenarios were chosen for the Skeena watershed modelling to capture a more extreme, worst case scenario (RCP 8.5) and a more likely, moderate mitigation scenario (RCP 4.5) of radiative forcing. Radiative forcing is the change in the net radiative flux of energy at the tropopause due to changes in atmospheric composition due to climate change or due to increased output from the sun (IPCC, 2014). Shown in added detail in Figure 1Ch1Appx within the appendix, as of 2011, the total radiative forcing relative to 1750 pre-industrial levels is 2.29 Wm^{-2} (IPCC, 2013). Unless substantial climate policy action and CO_2 emission negation are taken, RCP scenarios under 2.29 Wm^{-2} of radiative forcing by 2100 are becoming further out of reach each year. However, as evident in a decrease in global coal emissions from 2010-2019, high RCP's (such as RCP 8.5) with no climate policy and increasing coal emissions are also becoming highly unlikely and extreme scenarios (Hausfather and Peters, 2020). Instead, based on historical emissions, current policies, and pledged policies, moderate to weak mitigation RCP scenarios (such as 4.5, & 6.0) are more likely (Hausfather and Peters, 2020). However, there is always uncertainty in the prediction of future and possible climatic outcomes based on societal action. Depicted in Table 1Ch1, the RCP 8.5 worst case scenario increases emissions throughout the 21st century and does not reach relatively constant concentrations of CO_2 until after 2250 (IPCC,

2014). Alternatively, RCP 4.5 increases in emissions and radiative forcing in the middle of the century and stabilizes around 2100 reaching constant emission concentrations by 2150 (IPCC, 2014). Future research could pursue more RCP scenarios. However, due to the time constraints of this research, only two scenarios that were available for multiple GCM's on the PCIC data portal were pursued within this study.

1 Table 1Ch1: Descriptions of the RCPs used in the modelling of the Skeena watershed. Data was retrieved (unless cited otherwise) from IPCC (2013) and IPCC (2014).

RCP	Total radiative forcing	CH₄, N₂O, & CO₂ Concentrations	CO₂ Concentrations	Emission Curve Description	Relatively Constant Concentrations Achieved	Description
4.5	4.5 Wm^{-2} by the year 2100 relative to 1750	630 *ppm* reached by 2100	538 *ppm* reached by 2100	Emissions relatively stabilized by and after 2100.	After 2150	Stabilization of emissions scenario (IPCC, 2013); modest climate policy mitigation (Hausfather and Peters, 2020).
8.5	8.5 Wm^{-2} by the year 2100 relative to 1750	1313 *ppm* reached by 2100	936 *ppm* reached by 2100	Emissions increasing for some time after 2100.	After 2250	No climate policy action (NOAA, ND); very high greenhouse gas emissions (IPCC, 2013); worst case emission scenario (Hausfather and Peters, 2020).

HydroTrend 1981-2010 Validation

GCMs vs ECCC Climate Station Climatographic Comparison

All climate models tended to underpredict the temperature for the Skeena subbasins by up to 3.54 °C (Fig 1CH1 and Fig 2CH1). This may be due to the climate stations being located at lower elevations while the climate models are raster grids averaged over the entire basin, including multiple elevations. Climate model precipitation tended to be overpredicted for SB2 (Fig 2CH1) and underpredicted or overpredicted for SB1, depending on the model (Fig 1CH1) compared to the basin averaged ECCC climate station values. When examining mean annual values, the total mean from CANESM was the closest to the ECCC stations mean for temperature and the second closed for precipitation (Table 2Ch1). In other words, CANESM had the lowest RMSE for temperature (SB1 and SB2) and the lowest (SB2) or second lowest (SB1) for precipitation out of all the models (Table 2Ch1). In addition, CANESM had the highest r-squared values in all except one comparison to the ECCC stations (Figure 1Ch1 and 2Ch1). Therefore, for capturing both the ECCC mean and seasonal trend, CANESM performed the best out of the GCMs for the entire Skeena Watershed. The Access model had the closest overall mean for total precipitation (Table 2Ch1), but the model was less effective at consistently capturing the seasonal trend (it over or underpredicted some months) and therefore produced the lowest r-squared values (Figure 2Ch1 and 1Ch1). The Access model also had the highest RMSE for temperature and SB1 precipitation (Table 2Ch1). Meanwhile, CNRM captured the precipitation values more closely for SB1 and may therefore perform better at capturing the coastal SB1 portion of the watershed (Fig 1CH1). Although the CNRM model overpredicted the annual mean precipitation, it had the lowest RMSE error for SB1 (Table 2Ch1) as well as the highest r-squared (Fig 1CH1). This was the opposite for SB2 whereby, CNRM produced the highest RMSE for precipitation (Table 2Ch1). CNRM was the second-best performing of the models in capture temperature, after CANESM.

2 Table 2Ch1: 1981-2010 Annual Comparison of basin wide ECCC Station and GCM raster grids average precipitation and temperature.

SB1				
Climate Variable	ECCC Stations	CANESM	CNRM	Access
Precip (mm)	Annual Sum: 1980	Annual Sum: 1736 RMSE: 29.32	Annual Sum: 1929 RMSE: 20.4	Annual Sum: 1657 RMSE: 38.28
TMean (C)	Annual Mean: 7.04	Annual Mean: 4.33 RMSE: 2.85	Annual Mean: 3.87 RMSE: 3.28	Annual Mean: 3.5 RMSE: 3.7
SB2				
Climate Variable	ECCC Stations	CANESM	CNRM	Access
Precip (mm)	Annual Sum: 618	Annual Sum: 768 RMSE: 15.07	Annual Sum: 824 RMSE: 22.86	Annual Sum: 717 RMSE: 16.05

TMean (C)	Annual Mean: 3.65	Annual Mean: 2.14 RMSE:1.6	Annual Mean: 1.49 RMSE:2.3	Annual Mean: 1.1 RMSE:2.8

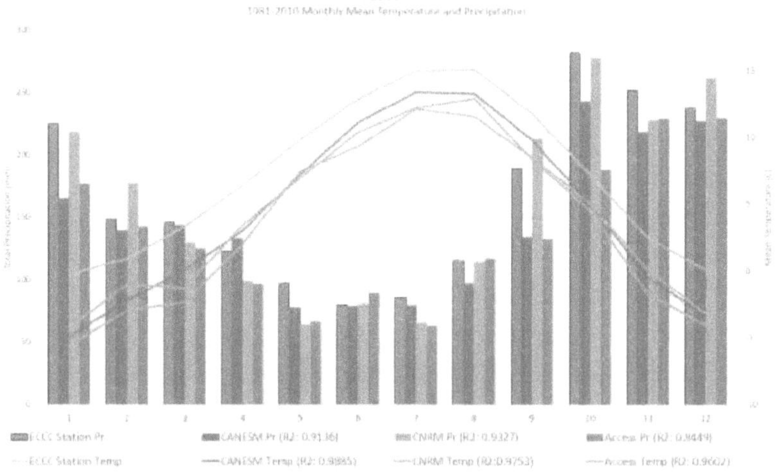

14 *Figure 1Ch1: SB1 Climate comparison*

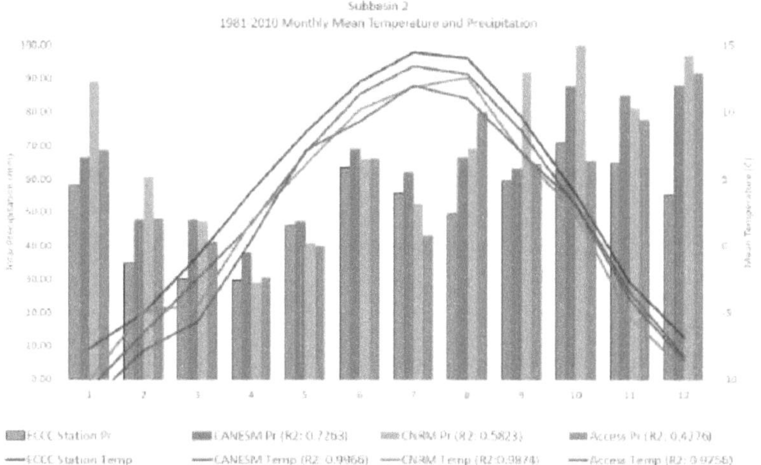

15 *Figure 2Ch1: SB2 Climate Comparison*

SB2 HydroTrend Results vs Usk Hydrometric Station
Mean Annual and Mean Monthly Discharge (Q)

SB2 model results of the watershed up to Usk are validated using the 1981-2010 monthly mean of daily Usk Hydrometric Station data (Figure 3Ch1). For the four validation runs, all input parameters are the same, except for basin-wide, monthly inputs of temperature and precipitation, from the ECCC climate stations, as well as the CANESM, Access, and CRNM Climate models (s. comparison in Section above). For the historical period, the RCP selection should not yet be at play as the climate changes induced by the RCPs apply to the future. Thus, only one simulation for each GCM is necessary for the historic period. Under future climate normals (2011-2040 onward) each GCM are presented under RCP 4.5 and 8.5 conditions. The validation runs cover the period 1981-2010 and are validated using mean annual and mean monthly averages of discharge at Usk during the same period. As shown in Table 3Ch1, HydroTrend using basin wide averaged ECCC station inputs produced the lowest mean annual percent difference (3.7%) to Usk mean annual discharge. All the GCM models overpredicted the discharge for SB2. The highest over predictions occur during the freshet (May-July) with up to ~2000 m^3s^{-1} (Figure 3Ch1). However, a t-test of the annual means shows that the Q of all the HydroTrend models are statistically able to accept the monthly means as equal to Usk Hydrometric station (Table 3Ch1).

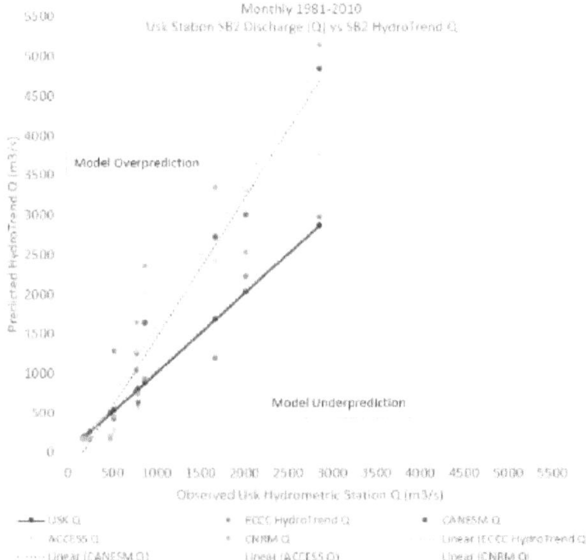

16 Figure 3Ch1: Usk Hydrometric Station Observed values (black) compared to HydroTrend outputs using various climate (ECCC vs GCM) inputs.
Values below the black Usk Hydrometric station line represent an underprediction of model results and values above the line represent model overprediction. Dotted lines represent linear trendlines for the model simulations. Twelve data points depicting the 30 year monthly mean appear on the graph per model simulation.

Mean Daily Suspended Sediment Concentration (SSC)

Observations of instantaneous suspended sediment concentration (SSC) at Usk are limited to 13 days in the years of 1988-1992, representing varied flow conditions (Q of 156 to 3290 ms^{-1} from Nov, Mar, Apr, May, June, Aug, and Sept (Figure 4Ch1)). Daily values of SSC were extracted for the same coverage (month and discharge) within the model runs are plotted against Usk values in Fig 4CH1. HydroTrend using the ECCC station averages for climate variables over predicted the 13 days of sediment concentration with an RMSE of 0.058 kgm^3 and a mean difference of 33.8 % (Table 3Ch1; Figure 4Ch1). All the GCM's underpredicted SSC (Table 3Ch1; Figure 4Ch1). The CNRM model underpredicted SSC the most of all the GCMs with a mean percent difference of 67.9 % respectively (Table 3Ch1). However, this did not translate into an underprediction of Qs (Table 3Ch1). Of the GCM models, the Access model appeared to follow the Usk station mean the most closely in terms of SB2SSC, followed by the

CANESM model. Due to the mean percent difference over 50 % for both Q and SSC, this research will not be moving forward with the CNRM model. Simulations of future scenarios in the sections below will use both the CANESM and Access models.

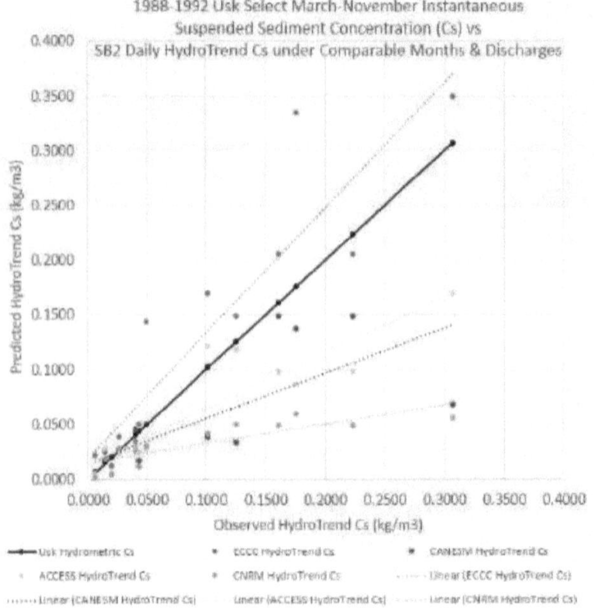

17 Figure 4Ch1: 13 Instantaneous Usk Hydrometric Station SSC compared to 13 daily SB2 HydroTrend SSC values.
HydroTrend SSC values were taken over the same period, month, and within 100 m^3s^{-1} of the Usk Hydrometric station discharge on the date the 13 samples were collected. The values below the black Usk station line represent underprediction by the HydroTrend model and values above represent overprediction.

3 *Table 3Ch1: Usk Hydrometric Station compared to HydroTrend Outputs.*

SB2					
Variable	Observed (Usk Station)	HydroTrend (ECCC Inputs)	HydroTrend (ACCESS Inputs)	HydroTrend (CANESM Inputs)	HydroTrend (CNRM Inputs)
1981-2010 Mean Annual Q	Mean: 903.6 m^3s^{-1}	Mean: 934.12 m^3s^{-1} RMSE: 290.85 m^3s^{-1} Mean Diff: 3.7% TStat: 0.18452 Two tailed critical values: 2.073873 T stat is between the + and – range of the critical value (alpha 0.05). Accept the mean as equivalent to Usk.	Mean: 1245.13 m^3s^{-1} RMSE: 630.42 m^3s^{-1} Mean Diff: 33.1% TStat: 0.72242 Two tailed critical value: 2.093024 T stat is between the + and – range of the critical value (alpha 0.05). Accept the mean as equivalent to Usk.	Mean: 1319.05 m^3s^{-1} RMSE: 757.17 m^3s^{-1} Mean Diff: 41.3% TStat: 0.80882 Two tailed critical value: 2.109816 T stat is between the + and – range of the critical value (alpha 0.05). Accept the mean as equivalent to Usk.	Mean: 1467.00 m^3s^{-1} RMSE: 967.32 m^3s^{-1} Mean Diff: 56.9% TStat: 1.03214 Two tailed critical value: 2.109816 T stat is between the + and – range of the critical value (alpha 0.05). Accept the mean as equivalent to Usk.
1988-1992 Instantaneous/ Daily SSC	Mean: 0.0996 kgm^{-3}	Mean: 0.1333 kgm^{-3} RMSE: 0.058 kgm^{-3} Mean Diff: 33.8%	Mean: 0.0640 kgm^{-3} RMSE: 0.061 kgm^{-3} Mean Diff: -35.8%	Mean: 0.0550 kgm^{-3} RMSE: 0.078 kgm^{-3} Mean Diff: -44.8%	Mean: 0.0323 kgm^{-3} RMSE: 0.100 kgm^{-3} Mean Diff: -67.6%
1981-2010 Mean Annual SSC	NA	Mean: 0.1225 kgm^{-3}	Mean: 0.0828 kgm^{-3}	Mean: 0.0704 kgm^{-3}	Mean: 0.0675 kgm^{-3}
1981-2010 Mean Annual Qs	NA	Mean: 205.96 kgs^{-1}	Mean: 215.04 kgs^{-1}	Mean: 219.27 kgs^{-1}	Mean: 222.99 kgs^{-1}
1981-2010 Mean Annual Qb	NA	Mean: 336.65 kgs^{-1}	Mean: 432.32 kgs^{-1}	Mean: 458.97 kgs^{-1}	Mean: 509.63 kgs^{-1}

SB1 HydroTrend Results (ECCC vs GCM Model Inputs)

No measured values are available to validate the SB1 HydroTrend simulations. However, a comparison of the HydroTrend simulations shows a similar trend to the SB2 results in that CNRM still outputs the highest Q and lowest SSC values relative to the other simulations using the other GCMs or ECCC climate station averages (Table 4Ch1). Although the CNRM climate variables match the ECCC climate data most closely for SB1 (Fig 1Ch1), Q, SSC, and Qb from HydroTrend simulations using CNRM inputs deviated the most from the HydroTrend simulation using the ECCC station data. This reaffirms the decision to not move ahead with the CNRM model when modelling the Skeena watershed up to 2100.

4 Table 4Ch1: SB1 HydroTrend Comparison

SB1					
Variable	Observed (Usk)	HydroTrend (ECCC Inputs)	HydroTrend (ACCESS)	HydroTrend (CANESM)	HydroTrend (CNRM)
1981-2010 Monthly Q	NA	$601.9\ m^3s^{-1}$	$699.9\ m^3s^{-1}$	$744.7\ m^3s^{-1}$	$834.3\ m^3s^{-1}$
1981-2010 Monthly SSC	NA	Mean: $0.3294\ kgm^{-3}$	Mean: $0.2502\ kgm^{-3}$	Mean: $0.2429\ kgm^{-3}$	Mean: $0.2118\ kgm^{-3}$
1981-2010 Monthly Qs	NA	Mean: $245.63\ kgs^{-1}$	Mean: $236.10\ kgs^{-1}$	Mean: $247.26\ kgs^{-1}$	Mean: $243.57\ kgs^{-1}$
1981-2010 Monthly Qb	NA	Mean: $76.38\ kgs^{-1}$	Mean: $88.81\ kgs^{-1}$	Mean: $94.5\ kgs^{-1}$	Mean: $105.87\ kgs^{-1}$

Combined HydroTrend vs CTDTu & Water Sample SSC at the Tidally Drowned River Mouth

ADCP transects, water samples, and CTDTu casts were collected in August and May, 2019 during the slack tide at Skeena River Downstream of Ecstall (SDE) (Fig SK1) location and can be used to compare to the combined (SB1 and SB2) HydroTrend SSC, Q, and Qs results (Table 5Ch1). CTDTu SSC rather than water sample SSC is used to capture a more proportionate average of the surface to depth throughout the water column. Low slack tide (moderate flow conditions) CTDTu casts and ADCP transects were used to compare to the model, as after ebb tide would be when the most riverine sediment would be settling throughout the water column. During the flood tide, more of the sediment may be from marine sources rather than riverine.

Simulation days with SB2 model results within 100 m^3 of the measured discharge at Usk and within the same season as the ADCP survey date on August 20, 2019 were selected to derive the combined SB1 and SB2 SSC, Qs, and Q. Simulation results of days that matched the ADCP survey condition discharge average to produce a model value to compare to the measured ADCP and CTDTu values at SDE. HydroTrend does not account for any tidal influence on the river and computes the final freshwater discharge and sediment load of the river. Since the SDE location is tidally drowned, the incoming oceanic water dilutes the incoming riverine sediment and freshwater producing lower concentrations. Thus, the ADCP discharge and CTDTu mean cross-section SSC is used to calculate sediment load, to compare to the HydroTrend suspended sediment load results. Refer to chapter three of the book for further information on the the calculation of discharge from the ADCP data and the calculation of SSC from the CTDTu. 2019 is situated within the GCM 2011-2040 climate normal. Therefore, the results of the GCM models validated against the CTDTu SSC will be from the 2011-2040 period with RCP 4.5 and 8.5 simulations available. CTDTu SSC is also presented against the historic normal 1981-2010 so that it can be compared with HydroTrend simulations using ECCC climate data that in all other parts of the validation were more accurate.

1981-2010 August HydroTrend values using ECCC climate inputs with SB2 Q within 100 m^3s^{-1} of the survey date discharge at Usk of 1260 m^3s^{-1} produced combined suspended sediment concentrations at the final river mouth of 0.038-1.06 kgm^{-3} with a mean of 0.313 kgm^{-3} (Table 5Ch1). The discharge at the river mouth (SB1 + SB2) computed by the model is 1834 m^3s^{-1} and the final Qs is 574 kgs^{-1}. SDE low slack tide CTDTu casts produced an average value for SSC of 0.062 kgm^{-3}. The discharge calculated for the SDE location from the ADCP cross-section at slack tide is 10434 m^3s^{-1} producing a Qs of 650 kgs^{-1} (Table 5Ch1). Thus, the model appears to under-estimate the measured SDE Qs by ~76.7 kgs^{-1}. However, the tidally influenced channel may have resuspension of bed material or an introduction of marine sediment that could increase the total input. With that in mind, the small deviation is encouraging. HydroTrend simulations run with GCM climate inputs were also compared to the measured Q and Qs values. Out of these group of simulations, the HydroTrend simulation using the Access model under RCP 8.5 for the 2011-2040 normal period matched the closest (135 kgs^{-1} lower) to SDE Qs results (Table 5Ch1).

5 Table 5Ch1: Comparison of Combined Subbasin HydroTrend SSC results (blue highlighted area) to the SDE River Mouth CTDTu SSC measurements (orange highlighted area).

RCP	Date	Data Method	Inputs	Discharge (m^3s^{-1})	SSC (kgm^{-3})	Qs (kgs^{-1})
NA	20-Aug-19	SDE Cross Channel at Slack Tide	ADCP & CTDTu	10434	0.062	650
NA	Aug 1981-2010	SB1+SB2 HydroTrend	ECCC	1830	0.313	574
4.5	Aug 2011-2040	HydroTrend	ACCESS	2190	0.199	436
8.5	Aug 2011-2040	HydroTrend	ACCESS	2650	0.1943	515
4.5	Aug 2011-2040	HydroTrend	CANESM	2110	0.164	345
8.5	Aug 2011-2040	HydroTrend	CANESM	2500	0.184	459

HydroTrend Results

1981-2010 HydroTrend (ECCC Inputs) Outputs for the Entire Skeena Watershed

Using ECCC stations as HydroTrend climatic inputs, Table 6Ch1 displays the annual riverine outputs for the entire watershed (including Ecstall tributary) of Q, Qs, and Qb. SB1, previously excluded from past estimates of Skeena Q and Qs, has the highest suspended sediment yield due to an immense contribution from glaciers and heavy precipitation across the basin. Thus, the subbasin combination approach highlights the importance of including the SB1 area downstream of Usk in estimates of Skeena River contributions to the estuary. SB1 and SB2 bedload's are provided below for reference but are not combined for the final Qb value of the watershed. Qb estimates are from simulations over the entire watershed area instead of in subbasins as described in the methods above. The combination approach for Qs and the average basin approach for Qb selected for the more conservative estimates of Qs and Qb produced by the model. For the 1981-2010 period, the suspended sediment load is higher than bedload for the watershed. 1981-2010 mean annual ratio is ~60-70% Qs to ~30-40% Qb of the total sediment load (S).

The SB1 and SB2 combination approach (50325.3 10^6 m^3yr^{-1}) produces discharge just under the past estimates of Skeena River discharge (55 503 (Benke and Cushing, 2009) – 68 023 (BC Ministry of Environment, 2011) 10^6 m^3yr^{-1}). However, since the SB2 sediment validation shows higher concentrations using the ECCC inputs (Figure 4Ch1), a slightly lower discharge could help to produce a more accurate total suspended load. Meanwhile, the HydroTrend simulations for the historical period in the one averaged basin approach (84510.5 10^6 m^3yr^{-1}), overestimated the past estimates of Skeena River discharge and is likely too high. For the one averaged basin approach, the overestimation is likely because averaging the highly different temperature and precipitation of the coast and interior basin, raises the overall temperature and precipitation of a greater area of the basin.

Discharge and suspended sediment estimated by the model from the SB1 and SB2 combination are highlighted in yellow along with bedload from the one-averaged basin (Table 6Ch1). An overestimation of discharge in the one-averaged basin approach would, in turn, overestimate bedload. In order to produce the most accurate estimate of total Skeena sediment load for 1981-2010, the 1-basin CsB outputted by the model multiplied by the SB1+SB2 combination Q produces the bedload highlighted in blue (Table 6Ch1). When adjusted to the combination discharge, the bedload is 4.36 10^9 kgyr^{-1} lower. The Blue highlighted analysis offers an alternate bedload for the detailed description of the historical sediment load. For simplicity, plots and data presented in the rest of the results into the future only use the bedload from the 1-basin approach without an adjustment for subbasin combination discharge. Although the 1-basin approach does overestimate discharge and therefore also bedload, it is still more applicable and produces lower values than combining the bedload of subbasins that would not be applicable. Ideally, further research is necessary to confirm the bedload component.

6 Table 6Ch1: Annual model outputs at the Skeena River Mouth for the period 1981-2010 using ECCC inputs. Yellow highlighted values are estimates using 1-basin averaged bedload, and SB1+Sb2 suspended load approaches combined. Blue highlighted values take 1-basin bedload concentrations and multiply them by the SB1+SB2 discharge to produce the final Qb that is added to Sb1+Sb2 Qs to produce S. Final Skeena values are in bold while the results for the whole Skeena Watershed 1-basin vs Sb1+Sb2 combination approaches are shown for reference. SB1 and SB2 results are also shown for reference and are presented in green in an attempt to separate them from the orange whole Skeena watershed results.

1981-2010	Area	Discharge (**Q**)	Suspended Sediment Concentration (**SSC**)	Suspended Sediment Load (**Qs**)	Bedload Concentration (CsB)	Bedload (**Qb**)	Total Sediment Load (S)	Total Yield
	km^2	$10^6\ m^3 yr^{-1}$	Kgm^{-3}	$10^9\ kgyr^{-1}$	Kgm^{-3}	$10^9\ kgyr^{-1}$	$10^9\ kgyr^{-1}$	$10^3\ kgkm^{-2}yr^{-1}$
Final Skeena Values (SB1+SB2 Qs and 1-basin Qb outputs)	**54409.2**	**50325.3**	**0.2937**	**14.78**	**0.1268**	**10.72**	**25.5**	**468.67**
Final Skeena Values (1-basin CsB multiplied by SB1+SB2 Q)	**54409.2**	**50325.3**	**0.2937**	**14.78 (SB1+SB2)**	**0.1268 (1 Basin)**	**6.36 (Q*CsB)**	**21.2**	**388.92**
Skeena Reference (1-Basin)	54409.2	84510.5 (too high)	0.3209	27.12 (too high)	0.1268	10.72		
Skeena Reference (Sb1+Sb2)	54409.2	50325.3	0.2937	14.78	Not applicable	Not applicable		
SB1	12052.6	19830.3	0.4064	8.06	0.1269	2.52	10.58	877.82
SB2	42361.0	30495.0	0.2200	6.71	0.3604	10.99	17.7	417.84

Shown in Figure 5Ch1 are 1981-2010 monthly mean Q and S from HydroTrend (ECCC inputs). The timing of the hydrograph is similar to that at Usk Hydrometric station (Figure SK3). However, there is a higher peak in Q and S in the fall months at the river mouth than at Usk station for this observation period. The freshet appears at the end of May or early June. Shown in Figure 5Ch1, Qs is highest during the spring freshet and fall discharge peaks. Qb is highest during the warmest months of the year. Ratios of Qs and Qb are 50% in the Jan-Apr and July-August non-freshet months.

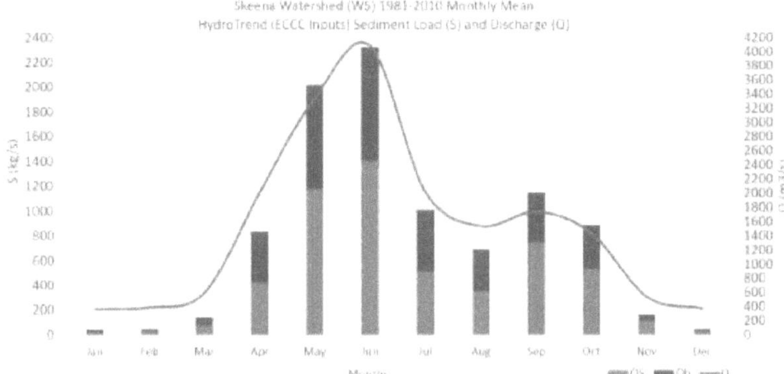

18 Figure 5Ch1: 1981-2010 Mean Monthly Qs, Q, and Qb at the Skeena River Mouth after the Ecstall tributary.

RCP 4.5 Scenario 30 Year 1981-2100 HydroTrend (GMC Inputs) Outputs

Four HydroTrend simulations for each GCM climate model cover the period 1981-2100 in 30 year intervals. Under RCP 4.5, for these four 30 year periods and using CANESM and Access inputs, mean annual Q increases during 2011-2040 and 2041-2070 before returning to levels slightly higher than during 1981-2010 by 2071-2100 (Table7Ch1). Qs initially increases from 1981-2010 to 2011-2041 before dropping below 1981-2010 levels by 2070-2100. Although the overall trends are similar between the CANESM and Access simulations, CANESM Q and Qs values and changes between periods tend to be higher than for the Access simulations.

Table 7Ch1 breaks down mean annual Qs and Q into the components simulated by the model (s. Table caption for details). As depicted in both the Access and CANESM simulations, ice melt is the highest in 2011-2041, contributing to higher discharge than in 1981-2010. However, ice melt reduces after 2041 while rainfall increases in 2041-2071 to produce comparable or larger discharge than in 2011-2040. By 2071-2100 higher rainfall and ice melt than during 1981-2010 produces a slightly higher total discharge. The model uses the same inputs for groundwater and subsurface stormflow regardless of the period simulated. While Qbase, therefore, remains constant, Qs varies slightly and increases over time. From 2011-2041, QsBQART derived from the basin wide BQART application increases, and QsG derived from glaciers decreases. QsBQART does not increase enough to replace the decrease in QsG, and therefore, the total Qs decreases by 2071-2100. Meanwhile, Qb increases over time for

CANESM simulations and increases and then decreases in the Access simulations. For both Access and CANESM simulations, final Qb is higher during 2071-2100 than in 1981-2010. Thus, the sources for Qs and Q shift under changing climate as well as the overall magnitude. However, any changes that occur initially are on the trend of being mitigated by the end of the century.

7 Table7Ch1: RCP 4.5 Annual Mean Q, Qs, and Qb results.
Qbase is discharge due to groundwater baseflow, QSS is due to subsurface storm flow, Qnival is snow melt, Qrain is rain surface runoff, and Qice is discharge due to ice melt. $Q_{S_{BQART}}$ is suspended sediment load derived from the BQART equation applied across the watershed, QsG is suspended sediment load from glacial melt, an Qb is bedload.

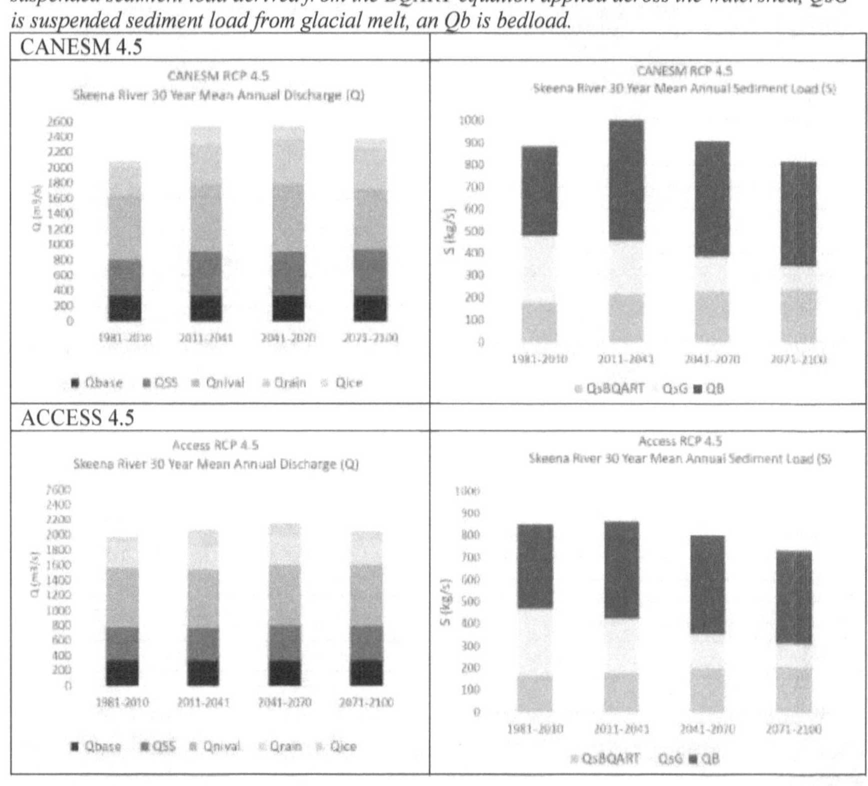

RCP 8.5 Scenario 30 Year 1981-2100 HydroTrend (GMC Inputs) Outputs

RCP 8.5 Qice reduces to equivalent or less than 1981-2010 amounts by 2071-2100. Qrain amounts increase and Qnival decrease by the 2071-2100 period. Thus, Q is highest in the middle of the century when Qice and high Qrain both occur. By 2071-2100 Q is higher or similar to 1981-2100 but composed of more Qrain and less Qice and Qnival. By 2070-2100, QsBQART is able to replace the Qs lost due to QsG. Thus, the suspended load of 1981-2100 and 2071-2100 are similar in amount but sourced differently. Qb increases by 2071-2100, producing a higher total S in comparison to 1981-2010. 2011-2041 has the highest Qs when glacial melt increases during the initial RCP 8.5 spike. In the 2041-2070 period, QsBQART is not yet able to replace the decrease in QsG and therefore, total Qs decreases.

8 Table 8Ch1: RCP 8.5 Annual Mean Q, Qs, and Qb results.
Qbase is discharge due to groundwater baseflow, QSS is due to subsurface storm flow, Qnival is snow melt, Qrain is rain surface runoff, and Qice is discharge due to ice melt. QsBQART is suspended sediment load derived from the BQART equation applied across the watershed, QsG is suspended sediment load from glacial melt, and Qb is bedload.

CANESM RCP 8.5	
Access 8.5	

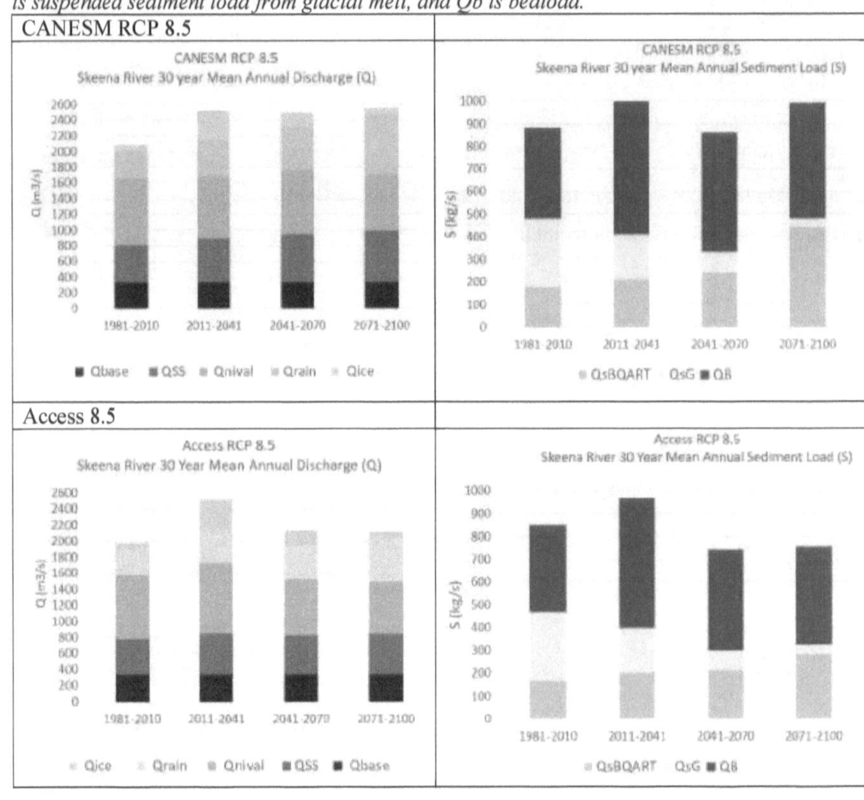

Shown in Table 8Ch1 and 7Ch1, HydroTrend simulations using the RCP 8.5 results differ from those using the RCP 4.5 scenario. The main differences between RCP 4.5 and 8.5 are the rapid glacial melt and associated loss of QsG as well as the rapidly increasing QsBQART in RCP 8.5 in the 2071-2100 period. For Q, the RCP 8.5 contribution of Qice is much higher during the 2011-2041 period than in the RCP 4.5 simulations. This is likely due to the rapid increase in temperature associated with the RCP 8.5 scenario. Shown in Figure 6Ch1, temperatures for the RCP 4.5 scenario begin to stabilize after 2040-2071. Thus, RCP 4.5 discharge begins to fall back to previous levels (Table 7Ch1). Increased temperature contributes to an increase in QsBQART, but it is not enough of an increase to compensate for the ELA retreat initiated earlier in the century. Thus, after a brief initial spike, sediment load especially suspended sediment load, decreases over time within the RCP 4.5 scenario. Meanwhile, increased precipitation, likely falling as rain rather than snow, as well as increasing temperature after the 2040-2071 period in RCP 8.5 (shown in Figure 7Ch1), drives a substantial increase in QsBQART (Table 8Ch1). Thus, under RCP 8.5, suspended sediment load returns to near 1981-2010 levels by 2070-2100 in a way not observed in RCP 4.5. This return to near 1981-2010 Qs levels by 2070-2100 is more apparent in the CANESM simulation than under the Access simulation. Furthermore, highlighted in Table 9Ch1, changes in Qs and Q sources are more drastic and rapid in the RCP 8.5 scenario than RCP 4.5, especially in the CANESM simulations.

9 Table 9Ch1: 1981-2010 compared to 2071-2100 under different RCP scenarios.

Change between the 1981-2010 to 1971-2100 periods	T (°C)	P (mm)	Qs (kgs^{-1})	Qb (kgs^{-1})	Q (m^3s^{-1})
ACCESS RCP 4.5	2.9	-0.06	-155.3	+37.5	+76.2
CANESM RCP 4.5	2.9	+8.98	-137.0	+66.6	+288.9
ACCESS RCP 8.5	5.4	+3.77	-145.1	+51.4	+139.5
CANESM RCP 8.5	5.9	+20.28	-1.3	+110.1	+463.2

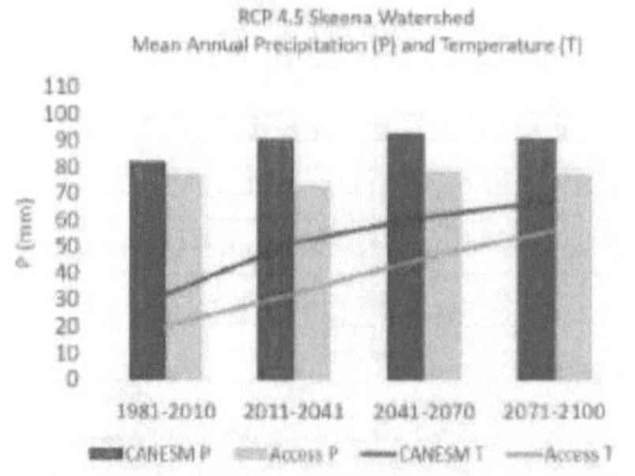

19 *Figure 6Ch1: Climate for the 1981-2100 period under RCP 4.5.*

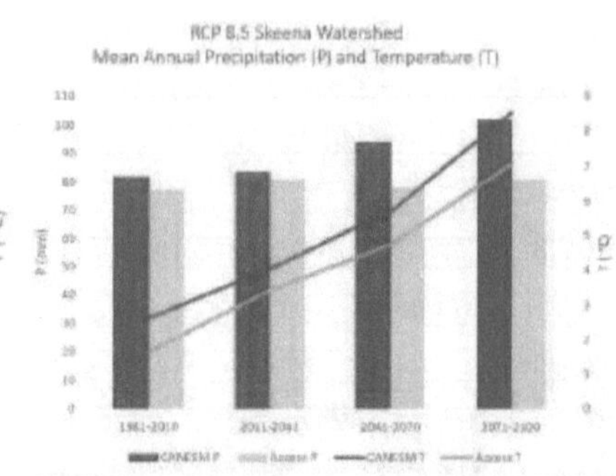

20 *Figure 7Ch1: Climate for the 1981-2100 period under RCP 8.5.*

Discussion

Skeena Sediment Load in Comparison to Past Estimates

Due to the lack of data available in SB1, previous estimations of discharge and the sediment load of the Skeena River during the 20th to early 21st century appear to focus on the river only up to Usk hydrometric station. This would underestimate the total discharge and sediment load of the Skeena River due to the importance of the SB1 contribution. Binda et al. (1986) estimate the sediment load of the Skeena to be 11 *million tyr⁻¹* for a watershed area of 42000 km^2 with a yield of 260 *tkm⁻²yr⁻¹* of sediment as described in Milliman and Syvitski (1992). The Binda et al. (1986) values were utilized in a global study on sediment flux by Milliman and Syvitski (1992). The total SB2 area up to Usk hydrometric station was calculated here as 42339.41 km^2 while the total Skeena watershed with SB1 amounts to 54 432 km^2. Therefore, it appears that the Binda et al. (1986) and Milliman and Syvitski (1992) estimates include only the Skeena watershed up to Usk Hydrometric station. By excluding SB1 area, past estimates of Skeena River sediment load may be underestimating S by up to 10.2-14.5 *million tyr⁻¹* when both Qs and Qb are considered. McLaren (2016) estimated the sediment load discharged from the Skeena to be 2-5 *million tyr⁻¹*. This would drastically underestimate the total sediment contribution of the Skeena by potentially over 20 *million tyr⁻¹* based on the HydroTrend results. Thus, the HydroTrend modelling results presented above highlight the importance of the SB1 portion of the watershed and estimate a higher riverine sediment contribution to the Skeena Estuary than past estimates.

As the second largest river entirely within British Columbia (~570 km), Suspended load is lower than the Fraser, but still quite high. The mean annual suspended load of the Fraser River from 1966-1986 is 17.1 million tonnes (metric) per year with a mean discharge of 3410 m^3s^{-1} at Mission (McLean *et al.*, 1999). Mission is approximately 85 *km* upstream of Sand Heads within the Fraser delta (Mikhailova, 2008). Based on the 1981-2010 HydroTrend results, the suspended load for the Skeena River is now estimated at 14.78 *M tyr^{-1}*. The Fraser is the longest river within British Columbia (~1375 km) with an active prograding delta (Williams and Roberts, 1988). However, there have been many cases when smaller or similar watersheds produce higher suspended sediment yields[28] (Richardson and Milner, 2010). Undoubtably based on the basin size and surrounding lithology, the HydroTrend results of this research agree with past literature (Mulder and Syvitski, 1995; Milliman and Syvitksi, 1992) that the Skeena River has a lower sediment contribution than the Fraser.

Effect of Changing Sediment Load

Based on the HydroTrend results, reductions in mudflats could occur into the 22nd century after an initial expansion during the glacial retreat of the 21st century due to a predicted increase and then decrease in sediment load. By 2071-2100, increases in temperature and precipitation cause an increase in basin-wide sediment export, but this is still lower than the past Qs due to a decreased contribution of Qs from glaciers. Predicted increases in discharge and decreases in Qs by 1971-2100 could result in reduced mudflat zones over long periods (Maan et al., 2018) within the upper Skeena Estuary. Increasing Q could increase the ratio of the export of fine material to the outer estuary rather than settling of Qs material along the upper estuary banks (Brouwer et al., 2018). However, despite a higher ratio of the incoming suspended sediment travelling to the seaward outer estuary, sedimentation in the outer estuary may not increase since the total incoming Qs from the river will decrease. In areas exposed to waves, less Qs deposition may allow for erosion without a subsequent replacement rate resulting in the degradation of exposed banks.

Higher Qb along with Q by 2071-2100 could coarsen the seabed leading into the estuary and promote the prevalence and size of seafloor transport features such as dunes along the delta plain. Thus, sand flats such as base sands fed from the increasing Qb may increase in size as

---

[28] For example, with a roughly comparable basin area, the Peace basin is reported to yield more than twice the sediment of the Fraser due to its surrounding lithology (Attard *et al.*, 2014).

well as an overall change to delta accretion and progradation. Schiefer et al. (2010) observed a lag between glacier retreat and heightened bedload due to the slower movement of coarse material through the river system. Therefore, once glaciers have retreated and the bedload has a longer time to equilibrate, the bedload may also decrease similar to the suspended load in the century earlier. It remains to be seen whether this increase in bedload transport leads to an increase in the rate of infilling sufficient to counteract or overcome the increase in accommodation space with rising sea level.

After an initial increase in sediment load during periods of glacier retreat in the mid-century, by 2071-2100, Q and Qs are predicted to decrease compared to 1981-2010 values, and this could potentially impact estuarine habitats. Decreasing the mudflat zones due to lower Qs (particularly the fine component) could potentially reduce the habitat for waterfowl and other intertidal species. A clearer water column with reduced SSC, on the other hand, could increase the health of eelgrass areas that support multiple juvenile fish species as well as benefit the survival of Skeena sponge reefs. However, a clearer water column could also expose juvenile salmon to predators (Lee *et al.*, 2005; Moore *et al.*, 1997). Further research such as modelling changes to the delta plain and estuarine zones could further validate theoretical conclusions as to how the estuary may change.

Additional Considerations

Although this research focused upon changes in climate within the next 100 years, changes in land use could also change export of the Skeena River. For example, increased mining, deforestation, or agriculture could all increase SSC concentrations (Schubel and Meade, 1977). Pasternack and Bush (1998) have observed through an analysis of multiple sediment cores from ten different tributaries, that watersheds with 40-50% of land under cultivation showed a doubling of sedimentation in comparison to the pre-settlement deposition rate. In contrast, the Huang He (Yellow) River, the second largest river in the world in terms of sediment load has shown a step-down decrease in sediment load from 1950-2005 due to the construction of dams, limiting historical flooding, sediment conservation practices, increased water consumption, and changes in precipitation due to climate change (Wang et al., 2007). Relative to other dammed and heavily mined rivers of the world under current conditions, all model runs of Skeena Qs used a low anthropogenic factor within the BQART Qs equation (see Table 1Ch1Appx). Further research could assess the effect of changes in land use within the watershed on the estuary.

Groundwater parameters were kept constant between model simulations when groundwater storage may also be impacted by changing climate. Further research could assess changes to groundwater parameters. Schiefer et al (2010) also caution estimating suspended sediment without long-term data validation due to potential variation in long term storage and sediment availability within the basin. However, storage and lithology are accounted for by the model within the BQART Qs equation as well as within hydraulic conductivity, subsurface storm flow, and groundwater storage. Further research could pursue long term sediment concentration and bedload observation program within the Skeena. However, with the current data available, a numerical model is an effective method to estimate riverine inputs to the estuary.

Chapter 2: Skeena Estuary and Delta Classification and Seabed Morphodynamics
Ch.2 Introduction

This chapter will establish an estuarine and delta sub-environment classification for the macrotidal, bedrock confined Skeena estuary and discuss how bedrock islands influence estuarine structure. In order to achieve the Skeena sub-environment classification, first, this chapter will describe the morphology and sedimentology of Skeena estuary and delta through the analysis of aerial imagery, multibeam bathymetry, grain size samples, and core-derived sedimentation rates. Second, conductivity, temperature, depth, and turbidity (CTDTu) casts derive water column salinity with depth. Rather than to discuss stratification, which will occur in detail in Chapter 3, CTDTU casts in Ch2 aid in the classification of different estuarine zones. Refer to Figure SK1 throughout this chapter for estuarine name and locations. More specifically, the chapter will address two objectives and hypotheses:

- Objective 1: Define the Skeena delta extent and discuss the impact of bedrock confinement on delta dynamics.

Hypothesis: A delta forms at the mouth of the Skeena Estuary that divides between bedrock passages. The extend of individual deltaic portions depends on the overall export of riverine sediment through the bedrock passages and introduces complexities that do not occur on a single delta deposit in an unconfined system.

- Objective 2: Identify different estuarine sub-environments and conclude how bedrock islands affect standard estuarine sub-environments.

Hypothesis: Bedrock confined estuaries will have higher variance than unconfined estuaries and will depart from previous classifications (e.g. Dalrymple et al. 1992).

Data Collection and Preparation
Bathymetry

Within the Skeena Estuary, multibeam bathymetry data were collected by the Canadian Hydrographic Service (CHS) using the Canadian Coast Guard Ship (CCGS) *Vector* over the past two decades primarily between 2005 to 2018 in water depths greater than 30 m. In 2019, CHS acquired LIDAR data over the temporally exposed intertidal area, Flora Bank. For other shallow, nearshore areas, such as the shallow delta plain approaching the river, without consistent high resolution multibeam coverage, CHS field sheets are available detailing navigational chart depth sounder data. In addition, in May and August 2019, in partnership with the Metlakatla First Nation the GECOS lab at the University of Victoria collected high resolution multibeam bathymetry along portions of the delta slope and nearshore of the estuary. GECOS Multibeam bathymetry was collected using a Norbit IWBMS Hydrographic Survey Sonar (Norbit, ND) and processed in Qimera Multibeam Data Processing and Analysis Software (Quality Positioning Services, 2020) using sound velocity profiles that were collected during the multibeam survey to account for changes in density with depth.

Seabed surface grab samples and sediment cores

Piston cores were collected by Natural Resources Canada (NRCan) from the CCGS *Vector* during cruises 2014007PGC, 2015002PGC, and 2016002PGC using standard NRCan piston coring methods (NRCan Expedition Database, 2019). Cruises 2017002PGC aboard CCGS *Vector* and 2015002PGC collected grab samples using a shipek grab sampler (NRCan Expedition Database, 2019). Cores were analyzed using a GeoTek© Multi-sensor Core Logger (MSCL-S) to record properties including density, magnetic susceptibility, p-wave velocity, and p-wave amplitude at the Pacific Geoscience Center (PGC). Cores were split and photographed using a high-resolution camera. Changes in stratigraphic layers within a core were described based on Munsell soil colours, texture, and structures based on visual analysis as well as using select grain size subsamples performed primarily by Pacific Geoscience Center (PGC) scientist, Kim Conway. Notes and diagrams from Kim Conway inform the descriptions on the Skeena cores within this chapter.

Approximately 270 seabed surface grab and sediment core samples retrieved between 1974-2017 within the Skeena Estuary were analyzed by NRCan for grain size distributions. Samples from 1980 and 1974 were sorted into percent gravel, sand, silt, and clay, using sieves by Luternauer (1976). In addition, 25 surface samples collected in 2014-2017 throughout the estuary, as well as additional samples at changes in sedimentary units throughout select radiocarbon dated cores, were subject to further grain size analysis using a Coulter laser diffraction instrument at the Bedford Institute of Oceanography. These samples have greater accuracy and precision (94 classes at ~0.2 PHI intervals) than the 1974 /1980 samples (4 classes). The grain size data was plotted and grouped based on the sorting, standard deviation, number of modes, as well as the occurrence of similar grain size (PHI) at the global maxima of the distribution using the method of moments (Folk and Ward, 1857). The Lax Kw'alaams First Nation shared 2474 grain size distributions collected from the Flora Bank area (McLaren, 2016). Additionally, CHS has approximately 20 000 points describing the seabed surface as bedrock, mud, sand, or gravel collected during multibeam surveying, grab, and core sampling within the Skeena Estuary over multiple decades. The grain size data were classified using the same convention as the CHS points and were together converted into Theisen polygons to produce a map of the seabed surface to complement the interpretation of the grain size distribution data.

Four cores were retrieved from a tidal flat along the banks of the tidal river just upstream of the Ecstall Tributary (Ecstall seen on Fig SK1). One of the tidal flat cores was subsampled and sent to Flett laboratories (Flett Research LTD., 2020) for Pb-210, Cs-137, Ra-226, and Be-7 analysis. Meanwhile, selected samples were extracted from 2014-2016 cores across the Skeena Estuary and were radiocarbon dated using Accelerator Mass Spectrometry (AMS) at the Keck Carbon Cycle AMS facility in Irvine California. OXCAL v.4.3.2 software by Bronk Ramsey (2017) was used to calibrate raw Carbon-14 dates. The Marine 13 calibration curve from Reimer et al. (2013) was applied to dates from shell samples with a standard marine reservoir correction of $\Delta R= 400$ years. The IntCal 13 atmospheric calibration curve from Reimer et al. (2013) was applied to dates from wood samples. For this research, present day, where the core is at 0 cm, is defined as 1950 (Enkin et al., 2013; Hamilton et al., 2015). When confronted with non-sequential dates down a core, the older date is disregarded as an outlier and the younger date retained for the calculation of sedimentation rates. All calibrated and raw Carbon-14 dates are in Table 1Ch2Appx within the Ch2 appendix along with the mean grain size at that sampling depth where available.

Water column data
Salinity data was collected in November 2018 from aboard the CCCG Vector using conductivity, temperature, depth (CTD) sensor. Casts were taken during the flood and ebb tide along a transect south of Telegraph Passage and west of Marcus Passage. Additional casts were collected in May and August 2019 from aboard the Metlakatla Spirit and Mission Point using a RBR Concerto CTDTu (conductivity, temperature, depth, and turbidity). Casts were collected in bedrock passages at three locations across the channel during the ebb, slack, and flood tide. The exact locations of these transects will be presented with the data in the results of this chapter and CTDTu sampling and stratification will be described in further detail within Ch3. For classifying

estuarine sub-environments, the upper 15 m of the water column and across channel casts were averaged to produce mean salinity values for each location. Averages were produced over the first 15 m to be comparable with the descriptions by Hoos (1975) who described the water column in the first 60 ft. Within November 2018 and May 2019 data, salinity varies between only ~2 PSU from the surface to depth, allowing this technique to be representative of salinities observed throughout the water column. Except for samples from August 2019 collected directly upstream of the Ecstall Tributary with freshwater dominance (~1PSU), August flows throughout the estuarine passages show variable salinity with depth. Therefore, August salinity data was omitted in this chapter. The range of salinity values thus utilized spans from low river inflows (November 2018, 309 m^3s^{-1} at Usk) to moderate river inflows (May 2019 627 m^3s^{-1} at Usk), with limited samples from moderately high inflows (August 2019,1545 m^3s^{-1} at Usk). Thus, for the estuarine sub-environment classification, general information on salinity under higher river inflows is also taken from Hoos (1975).

Additional Data Sources

Observations on Skeena Estuary morphology are drawn from Google Earth imagery, ESRI Digital Globe Imagery, PlanetScope satellite imagery, Google Earth imagery, and ShoreZone (2020) maps and aerial imagery. Cui et al (2017) digital geology, NRCAN and GECOS multibeam bathymetry, as well as geologic maps (Fulton, 1995; Turner et al., 2010) are also utilized.

Results

Skeena River and Estuary Morphology

Upstream of Terrace, the Skeena River meanders around a few large, forested islands, and forms point bars and cut banks typical of meandering river systems (Figure 1Ch2E). The course of the Skeena River is morphologically straight at Terrace where there is a relatively wide and flat area composed of glacial terrace deposits (Figure 1Ch2D). Downstream of Terrace, traversing through the Coast Mountains, the floodplain narrows (~ under 2 km) due to constraints of the harder intrusive and metamorphic bedrock (Cui et al., 2017). The river traverses through multiple channels anastomosing through several larger forested islands within a bedrock constrained flood plain between the surrounding Coast Mountains (Figure 1Ch2C). Approximately 13 km downstream of the furthest inland tidal influence at Kasiks River (Gottesfeld and Rabnett (2008), tidal limit), there is a visible transition in geomorphology as seen in satellite imagery (Figure 1Ch2). This transition is between an anastomosing river in a visible flood plain with large, forested islands exposed at all points of the tidal cycle to a tidally drowned flood plain with scarcely vegetated islands (Figure 1Ch2B). This "tidally drowned river" gradually transitions into a subaqueous delta with larger sand bars and mudflats with salt marshes at the banks, but no vegetated islands (Figure 1Ch2A). The Skeena River diverges into the bedrock confined passages of Inverness, Telegraph, and Marcus (Figure 1Ch2A). At the point of diversion, Marcus Passage has an average width of 2.7 km while Telegraph Passage is the widest at 3-3.3 km. Inverness Passage is the narrowest passage at around 1 km when measured from the bedrock and not the low tide banks. Telegraph Passage leads downstream into Grenville and Ogden Channel, whereas Marcus and Inverness Passages lead into Chatham Sound. Generally, passages become narrower in the middle reach of their length and wider approaching the tidal river mouth and oceanic Chatham Sound.

Satellite and aerial imagery show extensive tidal flats exposed at low tide along the channel banks in the "tidally drowned river" and subsequent estuary passages. The tidally drowned river and Inverness Passage appear to have the most considerable extent of tidal flats exposed at low tide. Satellite imagery shows that the tidal flats have remained relatively stable over the past twenty years. Sand bars appear to be parallel to the incoming and outgoing flow direction within Telegraph and Marcus Passage. Base Sand is the largest of these sand bars (Figure 1Ch2A).

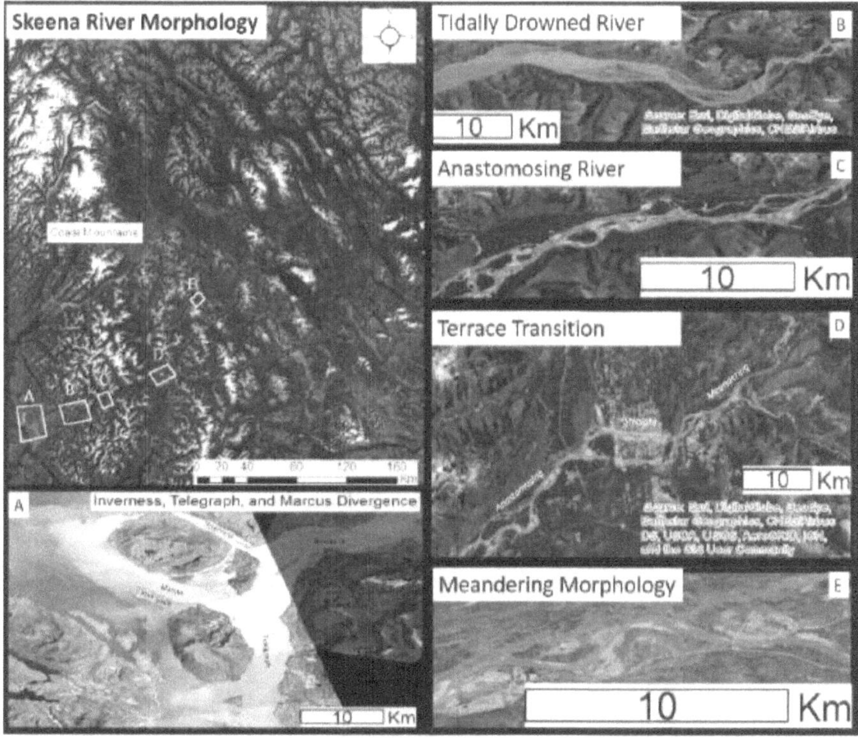

21 Figure 1Ch2: Morphological Progression of the Skeena River.
Insets have been labelled in white boxes on the watershed map with Figure A being the furthest
downstream and Figure E being the furthest upstream. Figure A tidal flats and sand bars have
been outlined in black to emphasize their presence. The Coast Mountain range is the last range
of mountains before reaching the ocean. The red line represents the furthest upstream tidal
extent described by Gottesfeld and Rabnett (2008).

Multibeam Bathymetry- Seabed Topography

      The bathymetric map displays the seabed within the estuary with a landward shallow
subaqueous plain transitioning into a deeper seaward basin (Figure 2Ch2). Leaving the Skeena
River seaward of Ecstall, there is a subaqueous plain at a mean depth of -5.38 m, a standard
deviation of 3.68 m, and a range from +7.17 – -12 m including the bars above the surface at low
tide (Figure 2Ch2). There is a deeper flow pathway scoured into the subaqueous plain with a
mean depth of -13.85 m, a standard deviation of 6.55 m, and a range of -7.4 to as deep as- 40 m
in select scour holes. Approximately 17 km downstream of the convergence of the Skeena with

its Ecstall tributary within Inverness Passage, 23-25 km downstream Marcus Passage, and 27-33 km downstream along Telegraph Passage the seabed transitions from the subaqueous plain into a deeper basin. The deeper basin seaward of the subaqueous plain still displays high variation in depth with shallower areas often present surrounding small islands. For example, a portion of Arthur Passage is shallower than the adjacent Malacca Passage and Ogden Channel (Fig 2Ch2). In the deeper basin of the estuary, the slope continues to deepen for up to eight kilometers from the landward subaqueous plain before reaching a seabed slope that is generally consistently under one percent slope change. The deepest depths within the estuary (down to a depth of approximately -250 m) are in the west or seaward leading areas in Chatham Sound and Ogden Channel (Fig. 2Ch2).

The slopes that lead out of Telegraph and Marcus Passages from the subaqueous plain transitions either gradually or abruptly into a deeper basin segment within a range of roughly 2 to 24% slope change (Figure 2Ch2). More specifically, the depth increases at the exits of Telegraph and Marcus Passages from approximately -3 to -12 meters down to between -100 to -161 m. The slopes between Marcus and Malacca/Arthur Passage were moderate or steep in comparison to other channels and ranged between roughly 4-7% slope change (Figure 2Ch2). Leading out of Telegraph Passage, the slope into Grenville Channel is gradual (4-2% slope change) with no subaqueous channels or failure deposits (Figure 2Ch2). Meanwhile, the Telegraph into Ogden Channel slope is either gradual (as low as 2% slope change) or abrupt (24% slope change) followed by subaqueous channels (Figure 2Ch2A). The more gradual slope of Ogden Channel that extends for up to 8 km aligns with the pathway of the deeper scour pathway on the landward subaqueous plain (Figure 2Ch2A). The more abrupt sections of the Ogden Channel slope occur surrounding multiple small islands and drops more rapidly from -12 m to below -100 in roughly 1000 to 1500 m from the subaqueous plain (Figure 2Ch2A). More abrupt slopes may be produced through a combination of a higher bedrock basement around islands or merely a more significant accumulation of sediment.

Inverness Passage transitions from a flat subaqueous plain bed (~7-8 m) to deeper water depths (12-32 m) with very irregular morphology at its NW end (Figure 2Ch2B). On the slope leading to deeper water, there are rhythmic, asymmetric bedforms up to ~5 m high. Seaward of this bedform field, there is a bulge (~1500 m wide) with blocky, disorganized morphology.

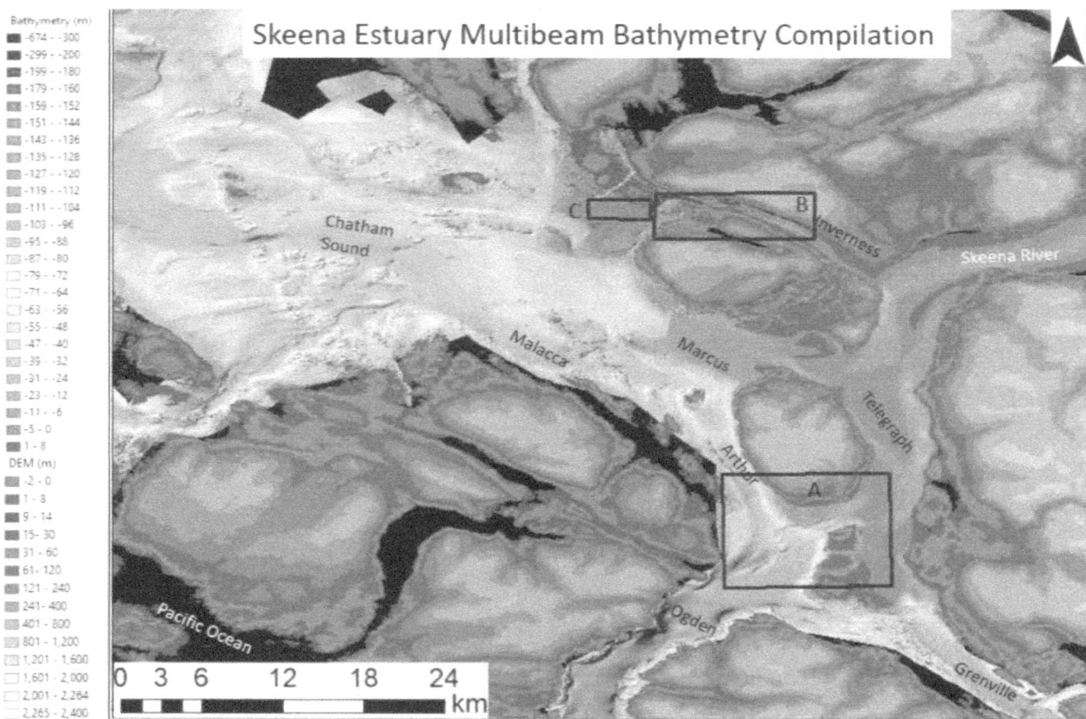

22 Figure 2Ch2: Combined Multibeam Bathymetry Image over the Skeena Estuary and Delta.
Blue square areas A-C will correlate with following multibeam bathymetry maps on specific features.

Submarine channels are only visible on the seabed in proximity to the subaqueous slope edge (Figure 2Ch2). More specifically, subaqueous channels leading away from the subaqueous plain slope are visible in the seafloor multibeam bathymetry between Telegraph Passage to Ogden Channel (Figure 2Ch2A) and between the entry of Marcus to Malacca Passage proximal to Base Sands (Figure 2Ch2). The Ogden Channel gradual slope produces the largest, in length and width, downslope subaqueous channels (Figure 2Ch2A). Portions of the abrupt slope are followed by smaller submarine channels (Figure 2Ch2A).

The slope transition from the subaqueous plain into the deeper basin is markedly different in Inverness Passage compared to Telegraph and Marcus (Figure 2Ch2B and 2Ch2A). Thus, to better interpreted these differences later, several initial interpretations have been made on the morphology of Inverness Passage (Figure 2Ch2B). The seafloor bulge in the multibeam bathymetry aligns with the location of the terrestrial slope failures (Figure 2Ch2B). Multiple sources describe past terrestrial landslide events within Inverness Passage, although there are no exact dates mentioned (Wishart, 2018; Gladys, 2001; Gulf of Georgia Cannery Society, 2020). Field notes during data collection also noted evidence of a past slope failure in the same area on the surrounding hillside that there is the slope failure deposit in Fig 2Ch2B. Thus, the bulge is interpreted as a subaqueous terrestrial landslide deposit on Figure 2Ch2B, leaving a mass of sediment on the seafloor. However, without further samples, the subaqueous terrestrial landslide deposit could also be a glacial lag deposit or deposit from when the river load reached further into the channel. Dunes are located in areas of channel deepening or around features that induce turbulence. This is consistent with the features observed on the slope transition from the subaqueous plain to the deep irregular bed, and thus, these features are labelled as dunes on Figure 2Ch2B. Finally, the irregular bed morphology in the deeper channel portion is labelled as bedrock on Figure 2Ch2B based on grain size samples described later in this chapters results and the fact that high tidal amplitude and velocities would likely rework sediment through the channel. Abrupt morphology observed at the seabed would need to withstand tidal flood and ebb velocities that would move sediment into a more gradual plain.

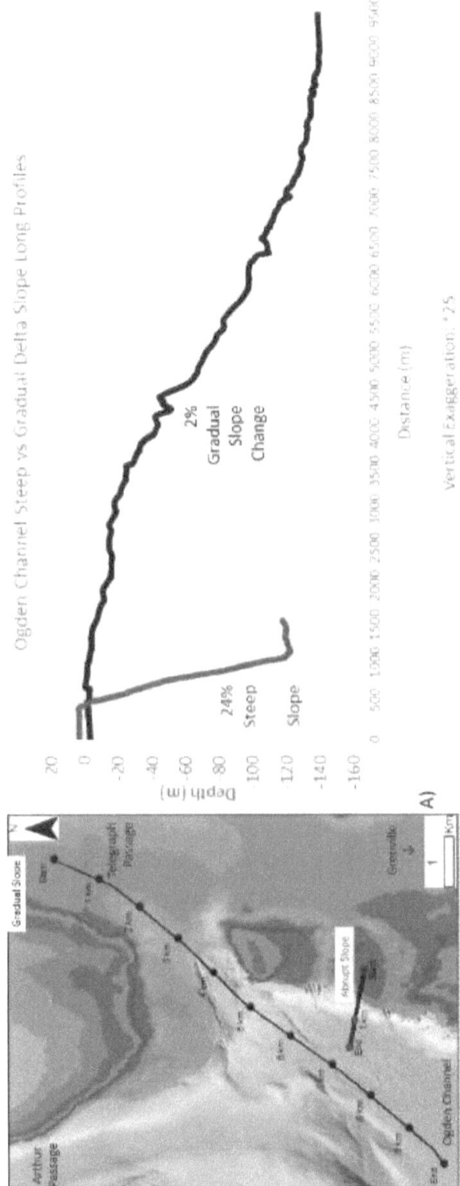

23 *Figure 2Ch2A: Telegraph Passage into Ogden Channel Slope.*
The black and blue lines on the bathymetry inset map (left) represent the pathway for the long profiles (right). Notice the subaqueous channels visible in the multibeam bathymetry of Ogden Channel. On the righthand graph, the same area on the graph is used to represent 20 m in depth and 500 m in length.

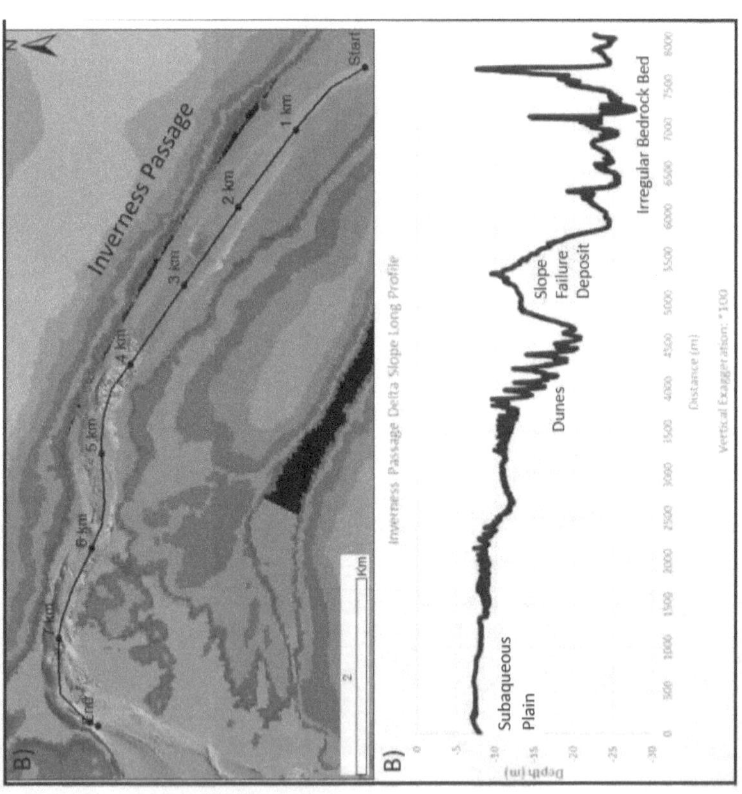

24 Figure 2Ch2B: Inverness Passage long Profile.
Black lines on bathymetry map (above map) represent long profile transects locations (below map). the same area on the graph is
used to represent 5 m in depth and 500 m in length.

Seabed Grain size Data

On the shallow subaqueous plain, grain size proportions consist nearly entirely of either gravel (upstream of Inverness Passage) or sand (88 – 100 % Weight Contribution, Figure 3Ch2); with sandy samples occurring most often. Finer deposits (dominantly silt or very fine sand dominant) accumulate along the banks and on the tidal flats. Leading seaward into Chatham Sound, Ogden Channel, and Grenville Channel, sand proportions decrease from the river through Marcus and Telegraph Passages. Eventually, in central Chatham Sound, Ogden Channel, and Grenville Channel, away from the subaqueous plain surface sediment consists of predominantly silt and clay with very little sand. However, in the outer mud dominated estuary, coarser grain sizes still occur around islands, banks, and inlets. Coarser grains also appear in the marine areas of the estuary further to the west within Chatham Sound (Figure 3Ch2).

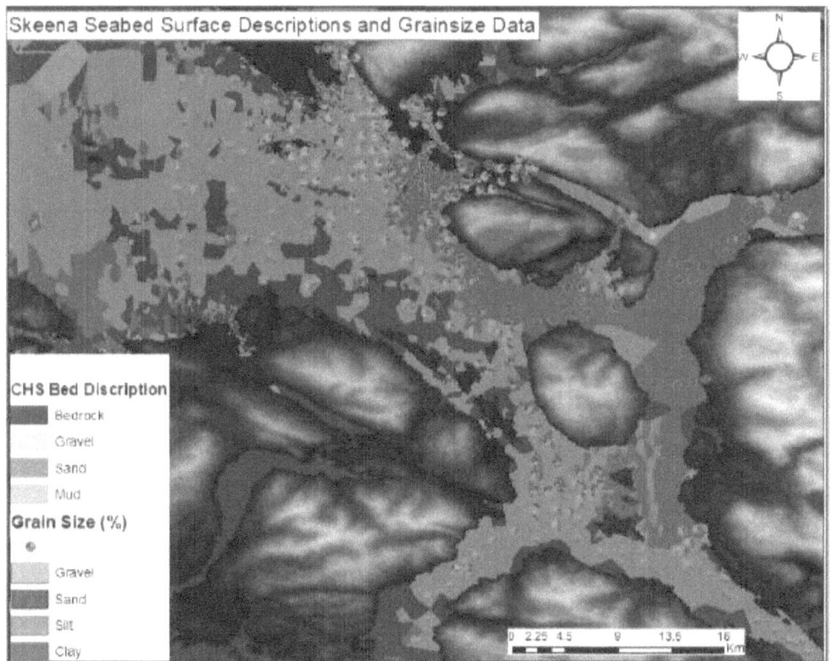

25 Figure 3Ch2: 1974-2017 Combined surface grain size samples within the Skeena Estuary. Only every 25th sample displayed due to the large amount of over 2000 samples over the Flora Bank area. The seabed Theisen polygon map was produced using samples from McLaren (2016), NRCAN, and CHS data.

Figure 4Ch2 displays selected grain size distributions along transects in greater detail. The distribution shows a fining trend in the dominant grain size from the oceanic and riverine entry points to the estuary (Transects A and C). However, distributions are very poorly sorted. Directly downstream of the morphological slope in Marcus Passage (surface sample of core 163), the distribution is a very poorly sorted and unimodal with a dominant mode in the fine sand to medium silt range and a long tail for finer grain sizes (Transect A on Figure 4Ch2 Map; Figure 4Ch2A). Further seaward, cores 164 and 165 display very poorly sorted, non-modal distributions across a breadth of grain sizes. The majority of the 164 and 165 sample falls within the coarse to medium silt range with sample 164 composed of slightly higher proportions of coarse material compared to 165 (Figure 4Ch2A). Surface samples from cores 163 and 164 display a fining trend in their mean grain size of 0.03467 mm (coarse silt) to 0.01606 mm (medium silt), respectively with distance away from the river mouth. Thus, sand is diminishing in amounts leading away from the morphological slope. Conversely, coarser grain sizes increase in proportions approaching the open ocean to the west (Transect C on 4Ch2 Map, Figure 4Ch2C). The dominant modes continue to coarsen in sample 167 to149 and finally results in a unimodal, very fine sand distribution with a long right-hand tail in the seaward sample 11 (Figure 4Ch2C).

The rate of fining leading away from the river and ocean is not uniform across Chatham Sound. This is not surprising given the irregular bathymetry and past glacial environment. However, trends are still present across the estuary. Coarser deposits tend to occur around the estuarine banks, especially near the passages connecting to the open ocean. At a similar distance from the subaqueous plain as sample 165, surface sample 2 displayed a very similar distribution (Transect B on Figure 4Ch2 Map, Figure 4Ch2B). However, sample 174 displayed a very different distribution similar to that of core 167 much further seaward into Chatham Sound. Both cores display a very poorly sorted distribution similar to other landward Chatham Sound samples but differ in the presence of a secondary mode in the fine sand zone that decreases at 0.08839 mm before increasing to a dominant mode in the silt ranges (Figure 4Ch2B). This decrease in the very fine sand range is visible in the surface samples in core 167, but not in cores 165 to 163 that are closer to the river mouth in Marcus Passage.

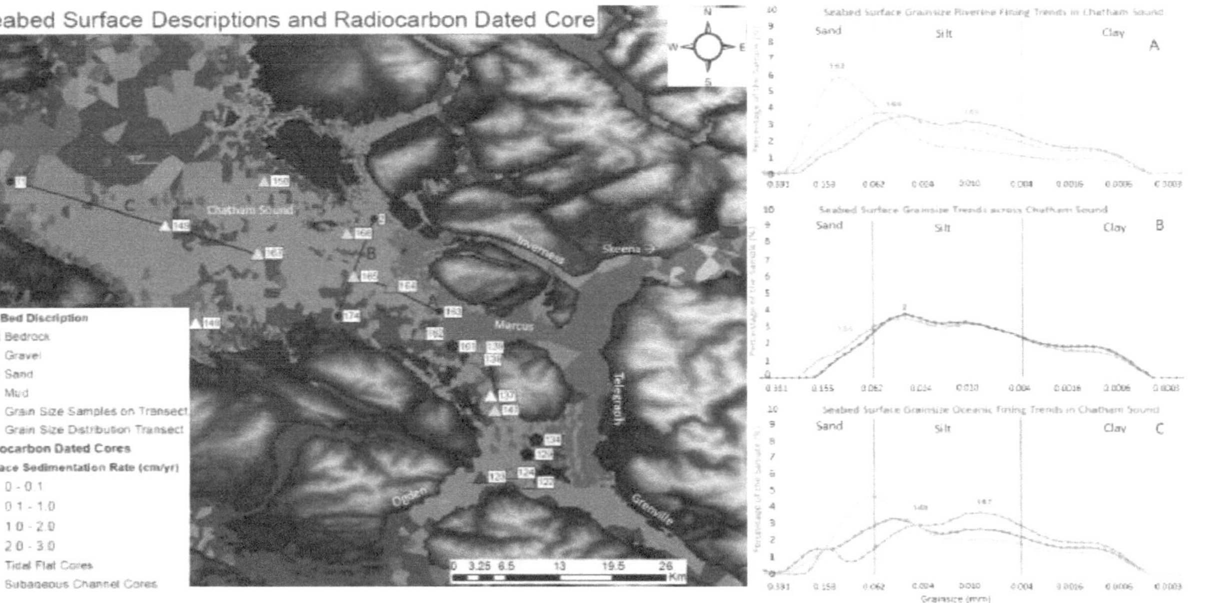

26 *Figure 4Ch2: High resolution surface grain size distribution transects, and Radiocarbon dated core locations. Radiocarbon dated and tidal core locations are also plotted on this map for reference in the next section of this report. The seabed Theisen polygon map was produced using samples from McLaren (2016), NRCAN, and CHS data.*

Marine Radiocarbon Dated Cores

Core Descriptions and Sedimentological Analysis

Cores that are proximal to Base Sands, Inner (landward) Chatham Sound, and Ogden Channel in areas without submarine channels, consistently display ungraded silt dominance with minor laminations and frequent bioturbation (122, 123, 124, 138, 139, 162, 164, 165, 166, and 167, locations shown on Figure 4Ch2). At the surface, mean grain size ranged from medium to fine silt. Distributions were all very poorly sorted especially at locations further seaward. At the surface, finer silt sediment is deposited in Ogden Channel (cores 122, 123, and 124) than at the exit of Marcus Passage (cores 162, 164, and 165). Colour throughout the cores is dominantly olive grey and more specifically ranged from 5Y 4/2, 5Y 3/2, and 5Y 4/3 on a Munsell soil colour chart. All cores tend to show bioturbation, buried shell fragments, weak laminations, and gas cracks. Some cores also have minor wood and sand lenses, especially proximal to Base Sands. Changes down core within silt dominant cores are minimal and thus can be considered one sedimentological unit. Minor changes downcore include occasional variations in color and minor, gradual changes in grain size. Core 165 in Figure 5Ch2 is a representative image of a Skeena silt dominant core.

Within submarine channels, cores display graded bedding and undergo large variation in grain size, colour, and texture (Figure 5Ch2, core 129, 134, and161). Generally, cores consist of sand and wood dominant layers of varying thickness (typically 2-30 cm) interbedded with mud layers. Mud layers are either bioturbated silt or laminated coarse muddy sand to sandy mud with buried organic matter. Sand layers varied between very fine sand to medium sand with visible proportions of mud and wood, to well mixed, very fine, fine, or medium massive sand deposits with minimal proportions of mud. Changes in grain size between layers of differing character occurred both gradually or abruptly. For example, core 161 is located in the distal end of an incised channel (Figure 2Ch2) with coarse sandy layers with significant woody debris interspersed with thick layers of bioturbated and burrowed silt. In core 161 (Figure 5Ch2) around 40 cm, the core transitions abruptly from medium sand (mean 0.35 mm) to fine silt (0.011 mm). Around 80 cm, the core gradually coarsens from fine sand (0.15 mm) to the medium sand (0.35 mm). An abrupt shift in grain size is indicative of an erosional or depositional event. Sand laminae and layers with sharp lower contacts throughout the core indicate higher energy flow

events. The woody debris capping such strata indicate density sorting of the sediments as the
energetic flows developed and are likely an indication of terrigenous and riverine input.

Silt Dominated vs Submarine Channel Core Images

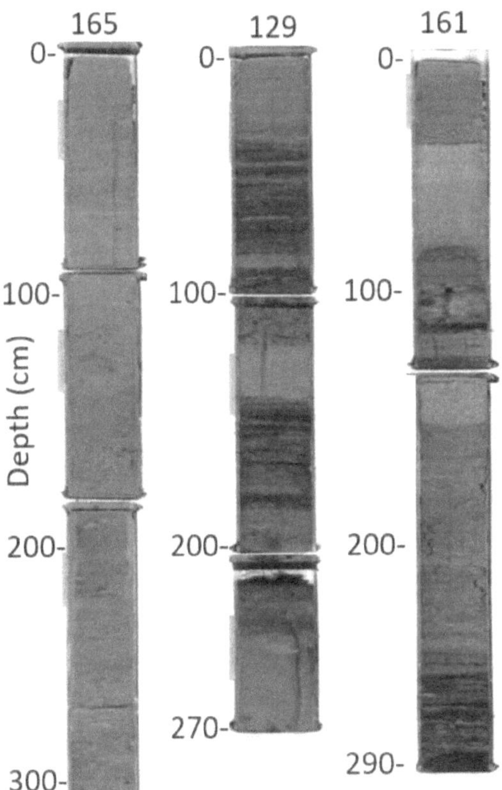

*27 Figure 5Ch2: Representative silt dominant (165) vs submarine channel cores (129 and 161).
Only the top 300 cm of the ~500 cm silt dominant core is shown.*

Outer Chatham Sound (seaward cores 148,149, and 150) and Arthur Passage (cores 137
and 147) are two locations within the estuary (Figure 4Ch2) where cores differ from the silt
dominant units seen elsewhere in the deeper portions of the estuary. Outer Chatham Sound Cores
are silt dominant at the surface but show different sedimentological units with depth in cores 149
and 150 (Figure 6Ch2). The bottom of core 149 and 150 consists of coarse sand and gravel

deposits, which transitions abruptly to silty clay. Shell fragments are present throughout the core except for the bottom. Core 149 and 150 coarsen up the core as sediment transitions from a dark grey, clayey mud to an olive green, bioturbated, very fine silty clay mud. Core 148 in Outer Chatham Sound is coarse (interbedded fine sand with olive grey, bioturbated, shelly mud) at the surface and finer with depth (dark grey to black, finely laminated to massive, very fine silt to clay) (Figure 6Ch2). Arthur Passage cores vary between very fine sand and silt sediment layers throughout the core or are similar to core 149 to 150 where sediment coarsen up the core from a clayey mud deposit to a well-mixed silty clay (Figure 6Ch2). Within the upper 300 cm of the core, core 149 (0.01053 mm at depth to 0.01408 mm at the surface) and 148 (0.00694 mm at depth to 0.04299 mm at the surface) exhibited a coarsening of silt sediment over time. At a similar distance from the river, the first 300 cm of core 150 exhibits a fining of silt over time, from 0.0152 mm at depth to 0.01038 mm at the surface. However, at depths greater than 300 cm, core 150 displays a clayey mud similar to what cores 149 and 148 display earlier in the core and closer to the surface.

Chronostratigraphy

Refer to Table 1Ch2Appx in the Ch2 Appendix for detailed information on lab numbers, raw radiocarbon dates, calibrated dates, and core details for chronostratigraphy. Calibration curves used to determine calibrated radiocarbon dates are also shown in Figure 1Ch2Appx-6Ch2Appx. Generally, single unit silt dominated and submarine channel cores (Figure 6Ch2) date within the last ~1000 years while Outer Chatham Sound cores date to the end Wisconsin glaciation (~13 300 BP) at the bottom of the core. More specifically, the oldest date measured at depth in the single unit silt dominant cores is 760 +/- 58 BP in core 165 at a depth of 443 cm (Table 1Ch2Appx; Figure 6Ch2). The youngest date at a similar depth in core 123 (489 cm) is 189 +/-72 BP. Submarine channel cores produce similar dates at depth to the silt dominant cores within the same area (Table 1Ch2Appx; Figure 6Ch2). Arthur passage core 137 dates are much older than the silt dominant core locations nearby (2220 +/- 66 BP at 130 cm depth)) and more similar to Outer Chatham Sound cores (956 +/- 62 BP in core 150 at 150 cm depth). However, Arthur Passage cores are shorter than those retrieved in Outer Chatham Sound and do therefore not date back to glacial times. Dates around 100 cm in the core are younger in core 150 than core 149 (~10 000 BP years younger) (Figure 6Ch2) potentially indicated increased sedimentation on the northern side of Chatham Sound.

Skeena Estuary Core Stratigraphy Diagrams with Calibrated Radiocarbon Dates

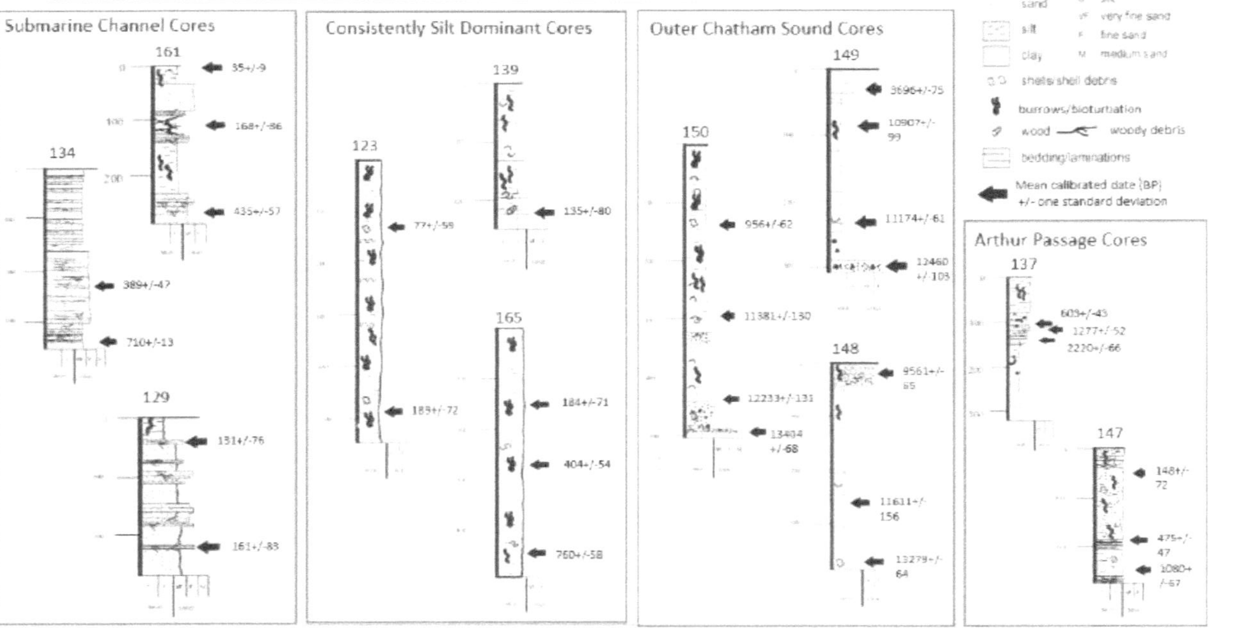

28 Figure 6Ch2: Representative core stratigraphy diagrams grouped by type (e.g.: consistent silt unit downcore) or area. Depth (cm)is on the y-axis and grain size classification is on the x-axis. The calibrated mean calendar date is displayed at the depth from which it was subsamples in the core along with the date's standard deviation. Refer to the legend in the righthand corner to interpret core diagrams. Initial Skeena core description notes, and diagrams were produced by Kim Conway (unpublished) and were modified to produce the figures above.

Sedimentation Rates

Upper core sedimentation rates from the first date retrieved in the core or from an average of sedimentation rates from dates within the last ~200 BP years are shown in Figure 4Ch2 and presented in greater detail in Table 1Ch2Appx as well as Figure 1Ch2Appx-6Ch2Appx in the Ch2 Appendix. Sedimentation rates in the estuary are under 0.1 $cmyr^{-1}$ in Outer Chatham Sound and above 1 and under 3 $cmyr^{-1}$ in Marcus and Ogden Passages (Figure 4Ch2). Except for Arthur Passage, sedimentation rates are generally higher (over 1 $cmyr^{-1}$ but up to 3 $cmyr^{-1}$) closer to the river and lower in the seaward portions of the estuary. Generally, sedimentation rates appear to decrease leading away from the riverine portions of the estuary.

Core Interpretations

13000 to 11 000 BP is considered deglaciation and 10 900 to 10 000 BP is the inundation period for the Skeena region (Clague, 1984). Thus, the fine clay deposits dating older than ~10 000 BP in Outer Chatham Sound are likely formed during the inundation and deglaciation of the Skeena region and are therefore described as glaciomarine. Similarly, the coarse gravel material older than ~ 12 500 BP at the bottom of the Outer Chatham Sound cores are likely relict glacial deposits that were buried during inundation (Figure 6Ch2). Wisconsin (>~10 000 BP) ages in the top 500 cm of Outer Chatham Sound cores could indicate ongoing reworking of sediments above the Wisconsin glacial and glaciomarine deposits through storm activity and low sedimentation. Introduction of fine material from the river is thought to contribute to the higher sedimentation and younger ages in the ~ 953+/-62 BP years of silt in Core 150 on the northern side of Chatham Sound. Again, as described previously, cores 149 and 148 on the other side of Chatham Sound have much older dates and different material within the top 100 - 200 cm of core.

During the past 1000 (maximum) years when silt-dominant cores were deposited, sea level is considered to be near its present level (See Clague (1984)). Sedimentation rates of 1-3 $cmyr^{-1}$ closer to the river within the estuary (Figure 4CH2) indicate that relict glacial features, if present within Ogden Channel, proximal to Base Sands, and leaving Marcus Passage into Chatham Sound, are likely buried and are not visible within a 500 cm core. A seaward fining trend, minor amounts of terrestrial wood imbedded in the cores, as well as recent dates (Figure 6Ch2) in these silt-dominant cores may be due to fine terrestrial material settling out from the river plume (visible in satellite imagery in Figure 1Ch2A).

The recent dates (past 710 +/-13 BP years) of buried material in cores of submarine channels (Figure 6Ch2) indicate that these are not relict glacial outwash channels preserved on the seafloor. Instead, heavy storms, high river freshets, and mass wasting from the nearby morphological slopes could be contributing to rapid and recent deposition of coarse and terrestrial material within the submarine channels. Further research would need to be conducted to examine whether these channels are currently active.

Based on the core descriptions and chronostratigraphy, core 137 in Arthur Passage appears to be out of the reach of the riverine plume, and likely situated over relict glaciomarine deposits resulting in older dates downcore than the surrounding Malacca and Ogden Passages. Core 147 has a unique stratigraphy that cannot be explained without additional information.

Intertidal Cores (Tidal Flat Cores)

Shown in Figure 7Ch2, tidal flat cores display mm to cm scale laminations of muddy sand layers interspersed between sandy mud. Locations of the tidal cores are shown in Figure 4Ch2 and 7Ch2. X-ray images and density measurements downcore show compaction. Denser areas within the core appear lighter due to decreased x-ray penetration (Algeo, 1994). This will occur as compaction decreases porosity and increases density downcore. However, lighter areas on the x-ray can also indicate coarser grain sizes (Grobe et al., 2017). Overall, cores coarsen with depth based on visual analysis. Lighter areas appear for rocks, gravel, and sand and dark areas appear where there are cracks in the core (black), wood and plant matter, and mud layers (dark grey). Density is the lowest for layers with high organic matter. This indicates that the x-ray pattern is more likely caused by grain size variation than solely due to compaction.

^{7}Be was not detectable in the first 3 cm of tidal flat core 1, suggesting that either the ^{7}Be level was too weak to be detected or that the sediments were not deposited in the last few months before the core was collected (Flett Research LTD., 2020). ^{137}CS was found in 0.47-0.12 DPM/g (disintegrations per minute per gram) throughout the core, indicating that the material was deposited after 1963. Measured at every cm downcore, ^{210}Pb analysis on one of the tidal flat cores indicates variable ^{210}Pb levels between 0.94 and 3.11 DPM/g. Due to this, variability, age modelling of the core was not possible. ^{210}Pb levels were significantly higher than ^{226}Ra, indicating that the background ^{210}Pb was not reached by the base of the core. This supports the

[137]Cs results that the sediments were deposited within, at maximum, the last 50 years (Flett Research LTD., 2020).

As described above, tidal flat core 1 displays fining over time with mm to cm laminations deposited within the last 50 years (Figure 7Ch2). Variations in grain size between laminations indicate changes in energy over the tidal flat, allowing different sizes of sediment to be deposited depending on the season or tidal cycle. Grain size samples at the bed in the center of the channel (see section on the seabed grain size data) just seaward of the tidal flats indicate gravel and sand deposits (Figure 3Ch2). Therefore, coarser deposits may source from the deeper portions of the morphological subaqueous plain during high flow events. Fine layers likely form under more quiescent flow conditions approaching the shallow and protected bank. Generally, the area of the estuary prior to the divergence into passages is well protected from waves (BC ShoreZone, 2020), potentially allowing for a greater accumulation on the flat. The fining over time displayed within the core may indicate a change in the hydrodynamic setting or expansion of the tidal flat allowing for less tidally remobilized delta sediment to reach the banks. As the tidal flat expands towards the channel center, fine sediments deposit on top of the delta bed. Changes in grain size of the material sourced from the river could also produce changes in grain size between laminations (if changes are seasonal as the discussion above with freshet flows) or fining over time (if changes are within the watershed over multiple years (e.g.: increased mining, forest fires, and more.).

Tidal Flat Cores Upstream from the Ecstall River

29 *Figure 7Ch2: Tidal Flat Core Locations and Imagery.*
Google maps 9/30/2018 Landsat satellite imagery with tidal flat core locations facing seaward (upper left). Aerial photograph of the tidal flat at low tide was retrieved from BC ShoreZone (1998-2000) imagery (lower left). High resolution imagery of split core 1 and Xray of core prior to being split were taken at the Pacific Geoscience Center (center). Lighter areas in the x-ray core appear where there are rocks, gravel, and sand (coarser grains). Darker areas are where there are higher proportions of silt and clay. The whole core was scanned through an MSCL-core logger at the Pacific Geoscience Center producing the gamma density. Dry bulk density was calculated at Flett labs (Flett Research LTD., 2020) from core subsamples (right).

Salinity

Under low flows (November 2018), the upper 15 m of water column had a range of 10-17 % freshwater. November transects across Chatham Sound (CS) show lower salinities (28.88 PSU) in the north-northeast (NNE) compared to the south-southwest (SSW) (31.46 PSU) (Figure 8Ch2). Tidal variance is negligible in the west (31.45-31.49 PSU) but increased towards the east of the transect (29.52 during ebb vs 30.8 PSU during the flood). Freshwater proportions are essentially the same in Grenville Channel (29.84- 30.12PSU) compared to Ogden (30.18- 30.90 PSU) based on the upper 15 m average of the water column (Figure 8Ch2). Grenville and Ogden Channel also have similar salinities than CS NNE.

SDE, Marcus, Telegraph, and Inverness Passages lie within the transition from freshwater dominance upstream of Ecstall to the saltwater dominance of Chatham Sound and Ogden Channel. These passages vary in salinity depending on the season, freshwater inflow, and tidal state. Telegraph and Marcus Passage are comparable at 29.08-29.38 PSU (17-16% freshwater) respectively under low flows (Figure 8Ch2), whereas Inverness Passage was fresher at 23.2 PSU (34% freshwater). Under lower flows (~693 m^3s^{-1} mean flow at Usk) (May 2019) salinity decreases (compared to the passages) to 18.25 PSU (48% freshwater) downstream of Ecstall (Figure 8Ch2). Upstream of Ecstall during moderate river inflows (August 2019) the water is essentially fresh and well-mixed (98% freshwater (0.86 PSU)). Thus, upstream of Ecstall is considered freshwater dominant under moderate river inflows. Further details on stratification and flow within the estuary will be discussed in Chapter 3.

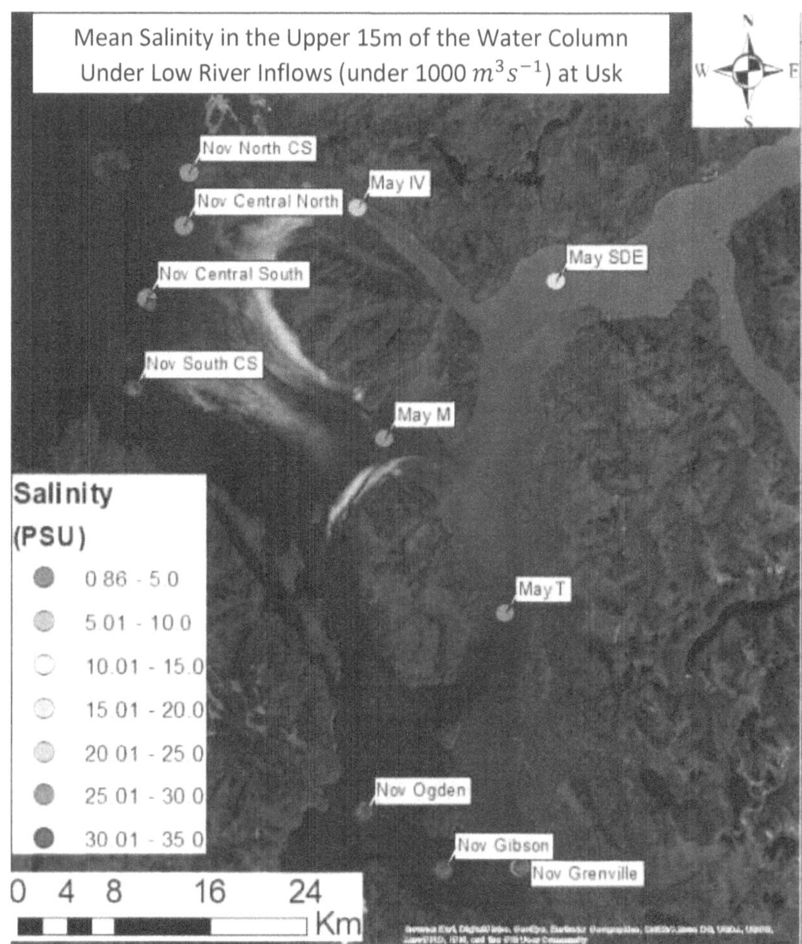

30 *Figure 8Ch2: Upper 15 m water column mean salinity from November and May campaigns. CS stands for Chatham Sound, IV for Inverness, M for Marcus, T for Telegraph, and SDE for Skeena Downstream of Ecstall ADCP survey locations. May values are averaged into one value for each channel location measured over a separate day over the ebb, slack, and flood cycle in the channel center, left, and right side. November samples were taken at one point of the tidal cycle with only select areas resampled 5-6 hours.*

93

Skeena Estuary and Delta Sub-Environments

The data presented above, as well as previous literature, were used to classify Skeena sub-environments (Figure 9Ch2). Refer to Table 1Ch2 for a summary of the classification criteria used. This approach draws upon estuarine facies classifications by Dalrymple et al. (1992) and Allen (1991) and applies them in a new context for the Skeena, using similar parameters such as bed grain size, morphologic features, and salinity to assign sub-environments. More specifically, the river dominated transport, estuarine convergence zone, and marine dominated transport estuarine zones are based on the tripartite structure described by Dalrymple et al. (1992) for macrotidal estuaries. Many fundamental principles of sedimentology and estuarine geomorphology such as the law of superposition, *Hjulström* curve relating sediment deposition and velocity, hydraulic geometry (e.g., Friedman and Sanders,1978); Allen et al., 1991), and Ritter et al.,2002) were critical for the distinction of Skeena sub-environments.

31 Figure 9Ch2: Skeena Estuary Sub-Environments and Features.
Further descriptions and distinctions between sub-environments will be detailed in the tables and text that follow.

10 Table 1Ch2: Skeena Estuary and Delta Sub-Environment Classification

Zone Name	Hydrodynamic Description	Morphological Description	Water Depth	Bed Grainsize	Sedimentation	Mean Salinity (upper 15 m)
1) Skeena River (SR)	Fluvial	-Anastomosing freshwater river	No Data	Sand to Gravel.	No Data.	Freshwater year-round (~100-99% freshwater).
2) Tidally Drowned Skeena River (TDSR)	Fluvial dominant but, tidally influenced area.	-Tidally drowned flood plain transitioning from partially vegetated to partially submerged islands. -Tidal flats line the banks and unvegetated bards are visible at low tide.	< 10 m	Sand to Gravel.	No Data in TDSR center. Tidal flat lining the banks show sedimentation over 1cm/yr.	Freshwater Dominant year-round (~over 50 to 99% freshwater).
3) Tidal Subaqueous Delta (TSD)	Transition from fluvial dominant delta to tidal estuary.	-Includes the delta plain, sand bars, and tidal scour channel as the river approaches the deeper estuary. -Tidal flats line the banks and unvegetated bards are visible at low tide.	< 40 m	-Sand & Gravel: >85%	No Data.	Transition from freshwater dominant to saltwater dominant depending on the location and season. (~98-15% freshwater)
4) Prodelta and Delta Slope (DSP)	Tidal estuary with fluvial influence on sediment transport.	-Includes an abrupt, gradual, and irregular delta slope leading into a more gradual Prodelta. -Tidal flats are visible in some inlets and sheltered bank areas however it is generally too deep for bars to appear at low tide.	< 161 m.	-Sand: 6-85% -Silt: 8-64% -Clay: 5-47%	Over ~1 cm/yr within the last ~200BP years.	Saltwater dominant year-round. (~Under 50% freshwater).
5) Transitional Prodelta (TP)	Tidal estuary with a seasonally fluvial influence. Aside from during peak river outflows, TP is very similar to EC.	-Prodelta fining interspersed with some very fine samples typical of the convergence zone.	< 255 m with a mean of 55 m.	Sand: 1-15% Silt: 49-64% Clay: 24-26%	Under ~1 cm/yr over the last ~200BP years.	Saltwater dominant. (~Under 20% freshwater).
6) Estuarine Convergence (EC)	Tidal estuary whereby fluvial and marine influence on transport are matched producing a low energy zone of fine material.	-Includes an area with samples low in sand and high in clay and silt.	< 391 m with a mean of 96 m.	Sand: 0-5% Silt: 30-75% Clay: 25-70%	Under ~1 cm/yr over the last ~200BP years.	Saltwater dominant. (~Under 20% freshwater).
7) Marine Estuary (ME)	Tidal estuary with a marine dominant influence.	-Seaward portion of the estuary with grainsize fining landward towards the convergence zone.	< 223.5 m with a mean of 91 m.	Gravel: 0-16% Sand: 1-67% Silt: 15-61% Clay: 9-53%	Under ~0.1 cm/yr over the last ~500 to 10 000 BP years.	Saltwater dominant. (~Under 20% freshwater).
8) Irregular Areas (Glacial Deposits) (IA)	Tidal estuary under varied conditions.	-Bed grainsizes are coarser than surrounding deposits.	Varied.	Varied, typically high in sand.	Under ~0.1 cm/yr over the last ~500 to 10 000 BP years.	Varied.

The Skeena River (SR)

The downstream end of the SR is defined based on a change in morphology from an anastomosing river with large, forested islands to a wide, tidally drowned flood plain with limited vegetated islands (Figure 9Ch2; Figure 1Ch2). This transition in morphology is located approximately 13 km downstream the tidal limit at the Kasiks River (Gottesfeld and Rabnett 2008). Within the SR, at Terrace and further upstream, banks are described as dominantly gravel (Turner *et al.*, 2010). The SR sub-environments zone is associated with freshwater and river dominated flow and is therefore not considered to be part of the estuary.

Tidally Drowned Skeena River (TDSR)

Since the Skeena Estuary is macrotidal, for all sub-environments onwards from TDSR, field data show that flow direction changes with the tidal cycle indication tide dominant flow. More details on flow will be provided in Chapter 3. The TDSR is characterized by shallow depths (-6 to +0.3 m) (based on multibeam bathymetry Figure 2Ch2) and a wide (1000 and 320 m) (google earth measurement Figure 1Ch2) passage downstream of the Skeena River sub-environment. Six km downstream of this marked transition to a tidally drowned river, the forested islands transition to sandy-gravel bars with limited vegetation and mark the end of the SR (Figure 1Ch2). The bars are only exposed at low tides (Figure 1Ch2). A tidal scour channel that incises to greater depth is visible in the multibeam bathymetry within the TDSR; however, it is less pronounced than the channel that emerges on the subaqueous plain (Figure 9Ch2 shown in yellow). The downstream end of the TDSR is defined downstream of the convergence with the Ecstall tributary when the Tidal Subaqueous Delta (TSD) begins.

Upstream (landward) of the confluence with Ecstall, CTDTu casts display vertically homogenous conditions under moderate river inflows (around 2000) producing a high water column averaged freshwater content (0.86 PSU or 97.5% freshwater shown further in Chapter 3). Under moderately low flow periods, Hoos (1975) in a transect leading away from Ecstall further seaward into the estuary and describe the lower end of the TDSR as having 70 to over 90% freshwater with proportions diminishing seaward. Based on these data, the TDSR is considered fluvial dominant yet, tidally influenced. Meanwhile, seaward at SDE, CTDTu casts show stratification under moderate river inflows and fall below 50% freshwater under low river inflows. Thus, seaward of the Ecstall convergence and prior to the observations of stratification and under 50% freshwater, the transition into the next sub-environment (TSD) is delineated. At the transition between TDSR and TSD, the river is 4 km wide and on average -3.7 m deep.

Tidal Subaqueous Delta (TSD, including Delta Plain, Tidal Scour Channel, and Sand Bars)
The subaqueous tidal delta sub-environment starts downstream of the TDSR and encompasses Inverness, Telegraph, and Marcus Passages. Evidence that the subaqueous plain can be considered a delta of fluvial origin is based on the morphology (gently sloping plain with incised channel stretching away from the river mouth), bed grain size distribution, and gentle slopes leading away from the platform. Cores downslope in the estuary also express high sedimentation rates with embedded terrestrial material indicating fluvial origin. Additionally, according to Friedman and Sanders (1978), deltas generally form at the mouths of rivers with suspended sediment concentrations exceeding 225 mgL^{-1}. Meanwhile, concentrations less than 160 mgL^{-1} are typical in estuaries without deltas. Areas between these concentrations are considered transitional. However, there are many exceptions due to waves, tides, discharge, depth, compaction, sea level rise, uplift, storm surge, and subsidence (Hutton and Syvitski, 2004; Friedman and Sanders, 1978; Seybold et al., 2007). 1981-2010 HydroTrend results using ECCC climate inputs described in Chapter 1, produced a combined 30 year mean SSC of 452 mgL^{-1} for the Skeena River. This indicates that the Skeena River has the sediment concentration to potentially form a delta based on SSC classification schemes should other conditions allow. The TSD zone is also well protected from waves, that can inhibit delta formation (Figure SK6). Thus, due to the evidence of grain size samples at the seabed, high riverine sediment input, and reduction of wave action in the TSD, it is likely that the subaqueous plain leading away from the TDSR is a delta plain.

The TSD is defined as all areas with over 85% sand and gravel at the bed (Fig 3Ch2) and shallow water depths leading away from the river mouth. The oceanward end is marked by a more rapid change in slope and a transition to less than 85 % sand at the bed (Figure 2Ch2 and Figure 3Ch2). The mean ratio of bed sediment in the TSD is approximately 99/1 sand and gravel/ mud, excluding the tidal flats. Within the TSD, a transition from freshwater to saltwater dominance occurs (Figure 8Ch2). At the SDE, freshwater ranges from 48 to 88% freshwater during low and moderately high flows, respectively. Telegraph and Marcus Passage ranged between 17 to 54% freshwater. TSD is also the only area of the estuary that showed high stratification under moderate river inflow conditions (Chapter 3). Based on these ranges, the

TSD ranges from 90-15% freshwater as the area transitions from fresh to saltwater dominance and varies with the seasons. According to Hoos (1975), the percent freshwater on the subaqueous delta plain and channel remain above 20 % in the first 15 m of the water column during both summer and freshet flow conditions (Figure SK7). Based on the rapid deposition of sand as the river slows and becomes dominated by the tide, the presence of tidal flats, as well as the change from freshwater to saltwater dominance, the subaqueous delta zone can be described as the transitional zone from fluvial dominance to the tidal estuary.

Tidal Flats and Salt Marsh Areas

Tidal flats and marsh areas occur around banks, islands, and inlets within the TDSR, TSD, and DSP sub-environments under the tidal influence, but sheltered from waves. The tidal flats are composed of mud and muddy sand and are visible in the satellite imagery at low tide (Figure 7Ch2; Figure 9Ch2). Tidal flats can also be seen in protected inlets and banks within the Prodelta (DSP) that are sheltered from waves and strong tidal currents. Sand deposition dominates the center of the bedrock passages and finer sediments are deposited in these areas where energy is lower and are reworked by the tides. Fining of sediments approaching the banks is a manifestation of the tidal influence on sedimentation (Freedman and Sanders, 1974) in the TSD sub-environments zone, whereas, the sand dominating the delta plain is primarily a product of the river bedload and ebb dominant flow (Conway,1996).

Prodelta and Delta Slope (DSP)

Broadly, the DSP sub-environment begins after the seabed transitions to less than 85% sand and continues as long as there is consistent fining in surface sediment leading away from the delta (Figure 3Ch2; Figure 9CH2). The mean depth in the Prodelta area is -50 m; however, may vary between -12 and -161 m (Figure 2Ch2). Sediment within the DSP has a wide range of grain sizes within surface samples. However, the zone can be defined as having under 50% clay and under 85% sand at the bed. More specifically, the samples from this zone ranged from 8-85 % sand, 8-64 % silt, and 5-47% clay (Figure 3Ch2). Sedimentation rates in this zone are over ~1 *cmyr^{-1}* and are the highest sampled within the deeper estuary (Figure 4Ch2).

All DSP and seaward sub-environments zones are considered saltwater dominant based on measurements in the upper 15 m of water. Inverness Passage (Figure 8Ch2) vary from 34 to 36 % freshwater under low to moderately high flows, respectively. Ogden and Grenville Channel vary between 12-15 % freshwater. Based on these values, the DSP is defined as having under 50 %, but over 10% freshwater. Based on the fining trend observed in the larger passages leading away from the delta, and saltwater dominance, the DSP zone is considered part of the tidal estuary that is fluvially influenced.

Telegraph and Marcus Passage Delta Slope

Immediately after diverging into Marcus Passage and Ogden Channel, the delta slope is attached to and potentially given support from the surrounding bedrock islands (s. the steep slope in Figure 2Ch2A). In some locations, this steep delta slope shows subsequent small, localized submarine channels. The scouring delta channel (Figure 9Ch2 in yellow) appears to extend downslope on the eastern side of Ogden Channel. In the lee of Kennedy Island, there is more of a gradual slope, and no continuation of the delta scour channel (Figure 9Ch2). The steep delta slope and extension of the delta scour channel into Ogden Channel may support the plunging flow necessary to form erosive currents under certain flow conditions that produce the large submarine channels downstream. The slope leading into Grenville Island is the most smooth and gradual. This area is also the most exposed to tidal reworking without an island lee sheltering the incoming or outgoing tidal flows. The delta scour channel is constrained to the southeastern side from which the tide propagates.

Inverness Passage Delta Slope

In Inverness Passage, the transition from TSD to DSP is defined at the seaward end of the gradient leading away from the delta platform that is covered by dunes (Figure 2Ch2B). Downstream the seabed is irregular and has been described as bedrock (McLaren. 2016; CHS). Therefore, the broad classification for DSP for bed grains does not apply here. Seabed grain size samples seaward of this bedrock portion and Northeast of Flora Bank indicate 6-19 % sand, consistent with the DSP classification. Inverness Passage is substantially narrower than the other bedrock passages. The lack of sediment cover in the seaward portion could either be due to ongoing scour by channelled tidal currents or due to a lack of deposition from fluvial or marine origins. The identification of dunes on the gradual slope does indicate strong tidal flows and reworking of the bed material.

Submarine Channels

Since the submarine channels are below an accumulation of sediment on the delta slope and a settling of the river plume as the riverine waters enter the deep prodelta (Figure 1Ch1A and

2Ch2), the submarine channels could be formed through turbidity current activity. Turbidity currents can erode the seafloor into channels and leave deposits of coarse material when they are active and finer material when inactive. Some of the fine material would erode when the channel is active producing abrupt and varied layering and varied sedimentation rates. This description is consistent with the submarine channel cores (Figure 5Ch2). Additionally, based on discharge and suspended load at Usk Hydrometric station (excluding SB1), the Skeena River was categorized into a moderately clean river category that may produce hyperpycnal plumes that induce turbidity currents with typical return periods between 100 and 1000 years (Mulder and Syvitski, 1995). More specifically, with an average Qs of 349 kgs^{-1} and Q of 920 m^3s^{-1} the Skeena was estimated to produce a hyperpycnal plume with a 2770 year return interval by Mulder and Syvitski (1995). However, Subaqueous channel cores with varying coarser grains that could be from turbidity current events are much younger than 2770 years (the oldest deposit is 710+/- 13 BP years shown in Figure 6Ch2). The inclusion of SB1 in discharge and suspended sediment load downstream of Usk (from Chapter 1, average total Skeena watershed Q is 1595.8 m^3s^{-1} and Qs is 468.8 kgs^{-1}) would likely shorten the return interval of hyperpycnal plume events for the Skeena. Thus, it could still be possible that the subaqueous channels form from hyperpycnal plumes. However, they are far from the river mouth (~30 km). Overall, further dating and analysis would be necessary for a more in-depth interpretation of the submarine channel cores.

Transitional Prodelta (TP)

The TP sub-environments zone is offshore the DSP sub-environments zone. It is defined based on a general decrease in % sand leading away from the delta slope (1 – 6 % sand) as well as an increase in finer sediments (clay and silt) that one would expect in the estuarine convergence zone (3Ch2; Figure 9Ch2). Occasionally, coarser sand deposits (diminishing proportions of 8-15% sand, Figure 3Ch2) occur with these finer deposits. This is likely because the TP zone is an area that may be dominated by fluvial transport and fining during freshet conditions and may transition into the convergence zone during lower flow conditions. Because of this, it is separated from the Estuarine Convergence (EC) sub-environments below. All samples show relatively high sedimentation rates (0.1-1 $cmyr^{-1}$). Sand ranges from 1-15 %t sand, 49-64 % silt, and 24-26 % clay (Figure 3CH2). Generally, lowest proportions of sand in this zone typically were found offshore Base Sands towards Prince Rupert (Figure 3Ch2). The position of the TP matches with the 6-10% to 10-20% ranges under freshet to average flow conditions of Hoos (1975) (Figure SK7) and falls into measurements of 12-17.5 % freshwater in Chatham Sound. Thus, the salinity range is ~6-20% for the TP zone. The hydrological condition can therefore be interpreted as a tidal estuary with seasonal fluvial influence.

Estuarine convergence (EC)

The estuarine convergence sub-environments zone was defined where sand proportions were under 5% consistently (Figure 9Ch2; Figure 3Ch2) with mean grain sizes in the fine to very fine silt range. There are noticeable contributions of clay that do not occur elsewhere in the estuary, Deposition of fine material indicates that the EC is a zone of low energy where fine sediments settle to the bottom and are poorly sorted. The bed coarsens towards the river and ocean on either side of the EC and it is located seaward of the TP sub-environment. The mean distribution for the convergence zone is 2 % sand, 49 %silt, and 49 % clay. The distribution ranged from 0-5 % sand, 30-75 % silt, and 25-70 % clay (Figure 3Ch2). Sedimentation rates in the EC range between 0.1 $cmyr^{-1}$ to 1 $cmyr^{-1}$ in the last ~150 years (e.g., core 122 and core 167 in Table 1Ch2Appx; Figure 4Ch2). The EC sub-environment is characterized by salinity values slightly lower than the TP sub-environment. Southwest Chatham Sound and Gibson Island measurements show 10-14% freshwater under low flows. However, Hoos (1975) found evidence of % freshwater less than 6%. Based on this, the percent freshwater ranges from roughly 20-1% within the EC zone. The main hydrological difference is that low energy is characteristic for the EC compared to the other tidal estuary zones.

Marine Estuary

The marine estuary (Figure 9Ch2) begins seaward of the EC where sand proportions exceed 5% at the seabed. A general seaward coarsening trend is characteristic for the ME and mirrors the increasing oceanic influence (Figure 4Ch2). After displaying a fining trend within the earlier zones, it is unlikely for fluvial transport energy to be high enough to transport sand material so far into Chatham Sound without eroding the finer material in the EC zone. Thus, sand deposits within this zone are considered to be of glaciomarine origin. The ME exhibits 0-16 % gravel, 1-67 % sand, 15-61 % silt, and 9-53 % clay at the seabed (Figure 3Ch2). More specifically, the mean distribution for the oceanic zone is 1 % gravel, 26 % sand, 41 % silt, and 32 % clay. Sedimentation rates over the last 411 to 9551 +/-65 BP years are under ~0.4 $cmyr^{-1}$ (e.g., cores 167, 148, 149, Table 1Ch2Appx). Such low sedimentation rates could also be indicative of limited transport. Sedimentation rates increase from west to east within the ME as well as with distance from the oceanic channel outlets. The marine estuary is saltwater dominant, with percent freshwater ranging between 15-1% in the upper 15 meters of the water column (Hoos, 1975; Fig Sk7). Based on these data, ME is classified as the marine dominant zone of the tidal estuary and has coarser material to the west near channels directly connected to the Pacific Ocean.

Irregular (Glacial) Areas (IA)

Irregular areas (Fig 9Ch2) are areas where sediment samples are coarser and sedimentation rates lower than in the surrounding areas (Fig 4Ch2). IA tended to occur near suspected glacial deposits, hexactinellid sponge reefs, fault lines, and slope failures. Hexactinellid sponge reefs tend to favor coarse glacial relic sediments (Shaw et al., 2018), which would confirm our IA classification for select areas of Chatham Sound. Cores 147, and 137 in Arthur Passage, with a dominant mode in the fine to very fine sand range and low sedimentation rates of under 0.4 *cmyr^{-1}* are located close to a fault line (erosional or lag deposits) (Cui et al., 2017). Glacial scour marks on the bedrock are visible in the multibeam bathymetry (Fig 2Ch2). Generally, greater variability in grain size distributions that deviated from surrounding trends occurred around islands and within inlets around margins (Fig 3Ch2).

Flora Bank

The largest of these IA zones is Flora Bank in the NE of the study area (9Ch2). Flora Bank is described as a stationary sand bank whereby the net effect of processes are equal on both marine and riverine sides of the bank (McLaren, 2016). Luternauer (1976) has described Flora Bank as a past or current deltaic deposit of the Skeena River based on his derived delta extent. However, according to McLaren (2016), all other grain size distributions in the Flora Bank area are said to be derived from one primary bimodal sand distribution of glacial till. Furthermore, the currents are too strong to allow the Skeena River to produce its signature on the seabed (McLaren, 2016). However, further research would be necessary to unpack the balanced energies concept for Flora Bank. Based on this new classification for Inverness Passage, the delta plain ends three-quarters of the way down the channel. Since the sediment has diminished in sand percentage prior to reaching Flora Bank, the area is not considered part of the delta and is therefore considered an irregular glacial deposit. Further data, such as core collection and dating, would be necessary to provide further insight into the past and origins of Flora Bank.

Discussion

The Physiography Classification of the Skeena Estuary and Delta

The Skeena Estuary and Delta has assumed its present state through a combination of changing sea level, macrotidal currents, and substantial river inputs within a bedrock constrained glacial valley. Estimation on Skeena River sediment load are given in Chapter 1. Further details on stratification and flow will be given in Chapter 3. The Skeena Estuary matches most closely to the Perillo (1995) (Shown in Figure LIT2) and Fairbridge (1980) definition of a fjard estuary. Although, the Skeena Estuary has shown certain elements of a tidally drowned river and certain features similar to Darlymple et al. (1992) tidally dominated coastal plain estuary. Fjard estuaries are tidally drowned glacial valleys that are low to moderately high in relief (~tens of meters) unlike their high relief (~hundreds of meters) fjord estuary counterparts (Perillo, 1995). Fjards also often exhibit multiple islands and moderate relief bedrock constrained channels. Macrotidal drowning and fluvial delta infilling traverse through and modify the Skeena fjard physiography. Even deep fjords such as the nearby Kitimat Fjord can produce fjord head deltas at river mouths (Shaw et al., 2017). Thus, with a moderate to low relief and given a substantial river input it would still be possible for a subaqueous delta to extend through the passages of a fjardic estuary.

The Evolution of the Skeena Estuary and Delta

The presence of glacial deposits is a characteristic feature of the Skeena River and Estuary. Alluvial and marine processes act to bury or rework the glacial material. 20-80 m drillholes along the Skeena River banks from Clague (1984) (Figure SK5) show alluvial deposits overtop of glacial mud in the lower reaches of the river. Surficial (2-5 m) Prodelta cores all indicate recent deposition (within the past ~760 BP years). Grain size fining at the surface of the seabed as well as the presence of terrigenous woody material within the deposit, indicate deposition through fluvial transport. It is only in the Outer Marine Estuary and in zones described as Irregular Areas where sedimentation is low, that glacial lag deposits are visible within the cores (Figure 6Ch2). Whether the Delta and Prodelta are overlaying glacial deposits at depth cannot be stated without additional, longer cores or seismic data collection. The fluvial deposition was likely initiated in this region immediately following the deglaciation period. The Fraser River in the South has been building its delta for the last 9,000 years (Williams and Roberts, 1988). How potential sea level regressions (Clague, 1984) would have affected the building of the delta is unclear. In the south, the Fraser showed continuous prograding throughout this time (Williams and Roberts, 1988).

Under present conditions, there are little areal floodplain deposits associated with the Skeena River delta. The glacially carved channel walls, as well as the rising sea level, combine to produce a large accommodation space that is deep rather than wide within the Skeena delta and tidally drowned river that must be infilled prior to sediments appearing above the water level. Estuarine basins infill when fluvial inputs are high by forming a delta or from deposits of marine and glacial sediments at the seabed brought landward by the tides that are not flushed away by river discharge (Friedman and Sanders, 1978). When sea level rises, the estuary or delta continues to infill the bed to raise the base level. Rising sea level in the Skeena Estuary continues to raise the accommodation space allowing further infilling of the current delta plain rather than prograding the delta front forward from present levels. For the Fraser River, the continued progradation is due in part to the large sediment load (~2.3 $M\ tyr^{-1}$ greater river suspended sediment load (not including the bedload that is vital for delta building) shown in Chapter 1) and the shallow basin geometry (Williams and Roberts, 1988). Thus, since there is such a large accommodation space and the sea level continues to rise, estuarine infilling and delta formation occurs and will likely continue sub-aqueously. High sedimentation rates in DSP areas near or downslope from the TSD, further indicate a dominantly submarine depositional environment in the Skeena areas leading away from the river (Figure 9Ch2; Figure 4Ch2).

The Impact of Bedrock Confinement on the Skeena Delta
The bedrock islands impact local exposure to tidal and riverine energies that allow for variability in the delta extend and slope characteristics. This variability may be due to the sheltering effect of bedrock islands, channelization of the tidal flow in delta scour channels, the amount of sediment that entered each channel, and the accommodation space for each channel (the basement bedrock depth and proportion of the channel left to infill). Each of the passages in Inverness, Marcus, Ogden, and Grenville have a different delta slope (Figure 2Ch2). When sheltered from the ebb and flood tidal currents due to the presence of bedrock islands, transport velocities are lower, allowing for sediment accumulation and gently sloping beds. When exposed to the ebb and flood tidal currents, slopes can be more gradual. Such a gradual slope can be seen leading into Grenville from Telegraph Passage (Figure 2Ch2A). Finally, in channels that have not been fully infilled or are sediment starved, the delta slope transitions to an irregular bedrock

bed. Although unconfined estuaries and deltas may have localized erosion and depositional areas, one would not expect to find areas that are sediment starved so close to the delta. Underlying bedrock can also produce very abrupt slopes that cannot be maintained by sediment under the current tidal regime.

On the Skeena delta plain, bedrock islands shelter passages from waves during winter storms (Hoos, 1975; ShoreZone, 2020). Sediment accumulation and delta building increase when waves suppress because there is shallower wave base and less basin reworking (Freedman and Sanders, 1978). Therefore, delta formation may expedite in narrow passages such as in Inverness Passage and in Telegraph Passage that are the most sheltered from incoming wave action by large bedrock islands. Generally, the delta plain has accumulated in the portion of the estuary labelled as protected (Figure SK6) whereas the delta slope is near the transition to a semi-protected wave area in Chatham Sound and Ogden Channel (BC ShoreZone, 1998-2000). Thus, variations in wave exposure induced by the presence of bedrock islands may have also played a role in controlling prograding. Bedrock channels constrain delta formation within a narrow channel and thereby favour progradation seaward rather than widening as seen in the typical delta (Δ) of an unconfined system. Wide receiving basins cause velocity to slow dropping coarser grains at its entry. If the receiving basin were narrow, higher velocities may persist and transport the material over larger distances. In the case of very narrow passages, this may result in material transported beyond the individual passage to the offshore. The Skeena delta extends furthest into Marcus (~24 *km* from TDRM) and Telegraph / Grenville Passages (~28 *km* from TDRM) and has limited extend in the narrow Inverness passage. The pathway to Ogdon Channel is narrow, and the delta slope occurs closer to the river mouth. However, sedimentation rates in the last 200 years at roughly the same distance from the river mouth (~33.5 *km* to core 123 compared ~32 *km* to core 165) are substantially higher in Ogden Channel (at ~2.5 *cmyr^{-1}* in core 123) compared to Marcus Passage (at ~0.8 *cmyr^{-1}* in core 165). This could be evidence that the narrow receiving basin of Ogden Channel concentrates sediment deposition further downstream within the Prodelta. Meanwhile, in the wider Chatham Sound sediment deposition is spread over a larger area.

Bedrock confinement increases rotational forces within the estuary along passage curvatures which tend to channel tidal flows and cause the formation of the incising tidal scour channel (Figure 9Ch2 in yellow). In an open, unconfined basin, rotational forces are less at play, and other forces such as Coriolis can dominate (e.g., quote). Confined curved bedrock passage can induce centrifugal forces on tidal inflow and outflow resulting in preferential erosion in the outer portions of the channel forming the tidal scour channel on the delta plain (Georgas and Blumberg, 2004). Pushing the tidal scour pathway into the outer portions of the channel would only increase under higher tidal flows and over time as the pathway deepens. In the seaward portion of the delta in Telegraph Passage, the scouring occurs more prominently along the south and southeastern portion of the channel before switching sides and moving closer to the center approaching the tidally drowned river mouth. The tide propagates from the south feeding into the delta scour channel. Opposite the scour channel, friction would increase towards the shallowing bed approaching the island's bank, slowing transport velocities, and at times allowing for the formation of tidal flats approaching the shallow banks.

Increasing the accumulation and subsequent failure of sediment due to the presence of bedrock islands could potentially lead to a higher downstream submarine channel activity. Furthermore, slope failures would further increase delta slope variability. Seaward of the Ogden Channel delta slope evidence of slope failures are the unsorted larger grain IA deposit (Figure 9Ch2 and 3Ch2) and submarine channels (Figure 2Ch2). Slope failures could have aided in producing the more gradual delta slope leading into Telegraph Passage from Ogden Channel while other parts of this delta slope remain abrupt (Figure 2Ch2). Submarine channels (Figure 2Ch2 seaward from Base Sands and Figure 2Ch2A on the Telegraph to Ogden delta slope) are in either localized areas coming off a steep slope (2Ch2 abrupt slope) or in a large set of channels within the prodelta transitioning gradually from the delta slope (2Ch2 gradual slope). However, further research would be necessary to confirm the relation of submarine channels to delta slope failure. Also, multiple factors contribute to the formation of submarine channels in fjords and along delta slopes aside from slope failures have not yet been fully considered. Regardless of the process that has caused the submarine channels, they are progressively inundated by rising sea levels.

The Impact of Bedrock Confinement on Estuarine Sub-Environments

Flemming (2011) describes how the convergence of opposing shores cause tidal energy to increase per unit width, whereas the added friction of the banks can cause energy loss to slow and reduce the tidal wave. Whether one dominates depends on the overall configuration and bed roughness. The bedrock channels are both rougher and deeper than sedimentary channels described by Flemming (2011). However, we do see that grain sizes are coarser at the mouth of Inverness Passage than a few samples upstream at the tidally drowned river mouth. Inverness Passage is a very narrow passage. Thus, this could indicate an acceleration of tidal and riverine energy at the divergence of the tidally drowned river into Inverness Passage. Seaward in Inverness Passage, bedrock is exposed at the seabed, which could also potentially indicate higher energy that does not allow for the settling of finer sediment. In comparison to the unconfined Darlymple et al. (1992) model, the bedrock-confined structure has focused both tidal and fluvial energy further seaward and landward and shifted the tripartite structure due to access to the marine and riverine areas.

Bedrock islands allow for a greater amount of IA zones and higher local variance between erosion and deposition within an estuary. The bedrock islands produce low energy areas where wave and riverine energy can be locally limited. For example, the IA in the western lee of Kennedy Island in Arthur Passage consists of coarser deposits and lower sedimentation rates than the surrounding areas. Due to this, past glacial deposits remain at the surface, such as has been postulated for Flora Bank (McLaren, 2016) and around the hexactinellid sponge reefs (Shaw et Al., 2018). In other equally dynamic, unconfined areas such deposits would likely be buried.

Bedrock islands within the Skeena Estuary cause a reduction in fetch, increase in friction, and thereby an increase in turbulence. They may also limit the effect of Coriolis deflection. In the largest basin of the estuary, Hoos (1975) observed a right hand or clockwise dominant presence of freshwater in Chatham Sound and attributed it to Coriolis force. In the outer estuary, higher sedimentation rates and larger grain sizes indicate the deposition of riverine sediment preferentially on the northeastern side of Chatham Sound (Figure 4Ch2; Table 1Ch2Appx). Salinity values were also higher in the NNE compared to SSW of Chatham Sound analog to observations by Hoos (1974) (Figure 8Ch2). These factors may be indicative of the Skeena plume 'hugging' the coastline and veering to the right of its motion (Figure 1Ch2A exiting Marcus Passage). However, Chatham Sound is much wider than the 5 km channels for which Coriolis deflection would shift direction with the tidal currents (Georgas and Blumberg, 2004), which would result in deflection to the southeastern side during the flood. Therefore, it is far more likely that the oceanic flooding from the northwest may be pushing riverine outflow to the eastern portion (landward side) of Chatham Sound and more deposition can occur. Thus, with more dominant forces such as tides, river currents, and waves, it is difficult to confirm whether Coriolis deflection is causing the higher sedimentation in the northeastern portion of Chatham Sound.

Based on the evidence presented above, bedrock confined tidal estuaries have greater variance from the tripartite structure of the Dalrymple et al. (1992) tidal estuarine classification (Figure LIT3). Although the Dalrymple et al. (1992) classification was not intended to address bedrock-confined estuaries, it is still useful to compare how the Skeena bedrock system differs from the Dalrymple al. (1992) models. The main difference is a shift in estuarine tripartite structure seaward due to the presence of the extensive subaqueous delta. Bedrock confinement constrains access to the estuary from the marine areas and created more protection from waves and storms for delta infilling and slanting estuarine zones based on marine and fluvial access points. There also appears to be a high number of IA zones and local variance within each estuarine facies zone that deviate from the original Dalrymple et al. (1992) descriptions. Furthermore, the Skeena River does not exhibit a straight, meandering, straight channel morphology approaching the estuary. Instead, the river is anastomosing, straight, and then diverted by a multitude of bedrock islands before reaching an open basin (Figure 1Ch2). This is because the Skeena River is a mountainous river that has been subject to glaciations that still influence transport and morphology along its reach and within the estuary. Despite these variations, the bedrock-confined sub-environments and Dalrymple's classification do have commonalities that are worthy of highlighting: Most prominent is the estuarine convergence zone whereby seabed grain size samples are still fining away from marine and riverine areas of the estuary as observed in unconfined areas.

Conclusions

Variance in the structure of the receiving basin, such as the presence of bedrock islands as opposed to an unconfined plain, can inhibit the formation of channel meanders along the river. Furthermore, bedrock islands and passages cause the system to deviate from typical estuarine classification and make it challenging to define estuary or deltaic conditions. Bedrock islands provide shelter from waves and increase friction towards the banks. Locally, tidal velocities may be higher due to channel confinement. This will alter transport energy and cause local variations in deposition/erosion. In the Skeena Estuary, a sandy-gravelly subaqueous plain exists indicative of a depositional delta structure. The feature has variable extends through the bedrock passages. More gradual delta slopes occur in the exposed areas of the channel where the tides can rework sediments. The tidal scour channel, noticeable throughout the delta plain, curvature and diversion are influenced by centrifugal forces that concentrate tidal current velocities. In the outer estuary, the direction of incoming tides and location of marine access channels force the dispersing riverine sediment in the opposite, lower to mixed energy areas of the estuary where fine sediments accumulate in the convergence zone. Overall, bedrock islands induce greater variance in delta slope production and estuarine structure through the disruption of marine and fluvial energy dissipation.

Chapter 3: Skeena Estuary Stratification, Flow, and Suspended Sediment Transport

Introduction

This chapter will identify the stratification and flow dynamics within different portions of the tidal cycle at various locations throughout the Skeena Estuary passages. Stratification and circulation are two metrics used to describe and categorize estuaries that have not yet been addressed within the Skeena. In a macrotidal area with high fluvial inputs and varied passage geometry due to bedrock confinement, it is previously unknown how different Skeena estuary passages will respond to ebb, flood, and slack tide under different river inflows. The nearshore Skeena estuary (under 40 m) was surveyed using an ADCP, CTDTu, and water sampling under moderate and low river inflows to assess stratification and flow. Further research is necessary to assess stratification and flow dynamics under high river inflows. The objectives and hypothesis of this chapter are as follows:

- Objective: Identify what areas of the estuary experience stratification under varied river inflow conditions. Identify the tidal break down of stratification and flow within a landward (prior to divergence) and seaward (post divergence of the estuary into multiple passages) estuarine passage.

Hypothesis: As freshwater spreads from the river source through different passages, salinity will increase and eventually dominate and dilute the water column. Therefore, stratification will be higher landward prior to the divergence of the estuary into multiple passages due to the higher proportion of freshwater to seawater. The river outflows will flow overtop of incoming flood flows (salt wedge) landward within the estuary, and this will eventually dissipate into well-mixed saline conditions further seaward. Furthermore, under similar conditions, stratification and flow will vary between passages due to their distance from the river source and bedrock divergence.

- Objective: Assess suspended material divergence into the seaward passages. Assess the impact of stratification and flow on suspended material flow dynamics over a tidal cycle within a landward and seaward passage.

Hypothesis: Landward suspended sediment loads will be higher due to the convergence of sediment mobilized during the flood and the prior divergence into various passages of sediment during the ebb. Sediment will be transported at the bed during the higher flood velocities and settle out from the surface during the ebb when tidal flows carry river outflows seaward.

Methods
Field Collection

Field data was collected on May 1-3, 2019 and August 19-23, 2019 in depths under 40 m aboard the Metlakatla Spirit using a WorkHorse Sentinel 600kHz ADCP, RBR Concerto CTDTu, and Niskin water sampler. Cross-stream transects at several sites (Figure 1Ch3) within the estuary were repeatedly surveyed with the ADCP for the duration of a tidal cycle. Refer to 1Ch3Appx in the Ch3 appendix for further survey details and dates. More specifically, four locations were surveyed in May and in August: Telegraph Passage (T), Skeena River Downstream of Ecstall (SDE), and Inverness Passage (IV). A fifth site, Marcus Passage (M) was surveyed in May but was not accessible due to high waves in August. Skeena River upstream of Ecstall (SUE) was surveyed instead of Marcus in August. At each site, survey transects alternated between having only the ADCP running as the survey vessel travelled straight across the channel to deploying the CTDTu and Niskin sampler at select locations while running the ADCP. CTDTu casts were deployed to the bottom depth at three locations across the channel, and water samples were taken at the surface, middle, and bottom of the water column at the same CTDTU cast locations across the channel.

Throughout the results of this chapter, the data presented will specify the May (low river inflows from flood to ebb) or August (moderate river inflows from ebb to flood) flow conditions. August survey conditions occurred during the ebb, slack, and flood tide periods surrounding low water level. In contrast, May survey conditions occurred during the flood, slack, and ebb tide periods surrounding high water levels (Table 1Ch3Appx). River inflows in August ranged between 1040 and 2080 $m^3 s^{-1}$ at Usk hydrometric station over the survey week (Table 1Ch3Appx) and are referred to as 'moderate river inflows' based on the Usk hydrometric station 100 year mean hydrograph (Figure SK3). In May, Usk hydrometric station inflows ranged from 580-690 $m^3 s^{-1}$ (Table 1Ch3Appx) and are referred to as low river inflows. Flow conditions were similar (within ~100 $m^3 s^{-1}$ of river inflow) between various estuarine sites surveyed on different days in May, whereas the river inflow was varied (within ~1000 $m^3 s^{-1}$ of river inflow) during August survey dates.

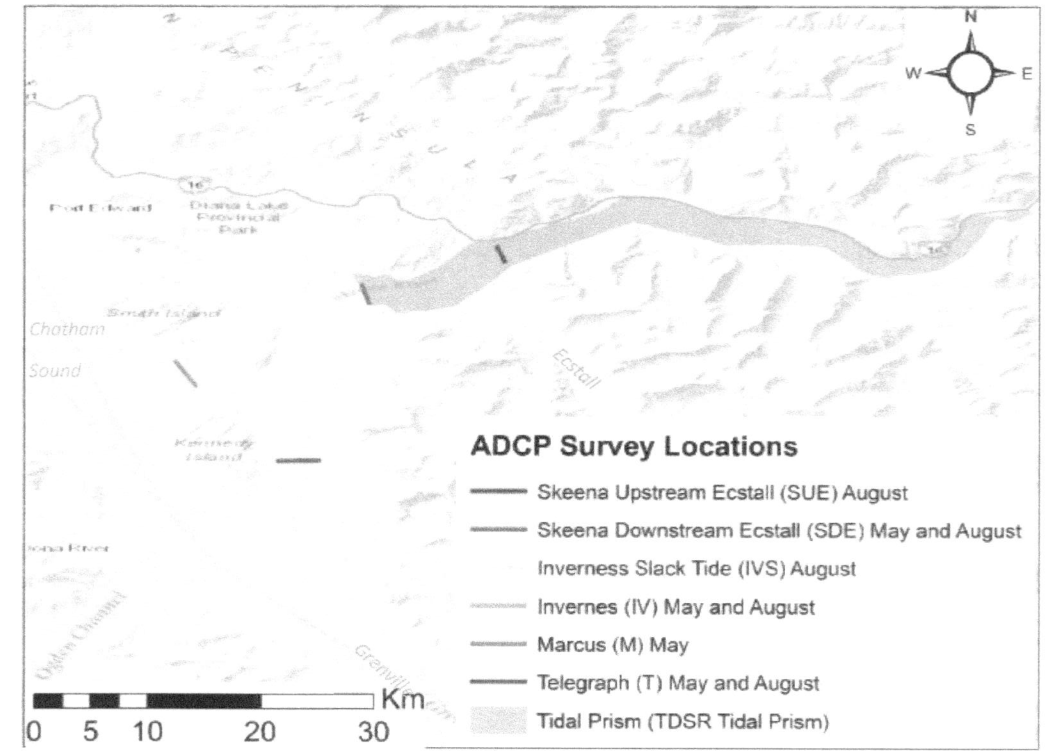

32 Figure 1Ch3: *May and August Nearshore ADCP transect locations across the estuary.*

CTDTu Turbidity Calibration

Water samples were filtered through pretreated Whatman GF/F 0.47 47 μm glass-fiber filters (47mm). Water sampling pre and post fieldwork treatment were adapted from the protocol of Röttgers et al. (2014) as well as the University Victoria Spectral Lab described in Phillips and Costa (2017). Water samplers were flushed with 100 ml of deionized water to remove debris and salt retention from the filter, stored in a desiccator, dried overnight at 60°C, and weighed three times on a μg precision scale. Filters were then burned at 500°C for 1 hr. and slowly cooled down to room temperature to remove the organic carbon and determine the inorganic carbon content. This combustion of organic matter is referred to as the loss on ignition (LoI) method (Schartau et al., 2019). Select water samples in May at the Skeena River downstream of Ecstall (SDE) used different Millipore HA mixed fiber filters (0.45 μm and 47mm) to determine suspended sediment grain sizes for the ADCP sediment inversion. The same treatment method was used, but filters were not burnt, as this would destroy the mixed fiber filters. Suspended sediment grain sizes were determined through a Coulter Counter Multi-sizer using a 30 μm and 200 μm aperture tube at the Bedford Institute of Oceanography.

The CTDTu measured or calculated, turbidity, conductivity, temperature, depth, salinity, pressure, and density. The starting time of the cast matches the CTDTU cast to the ADCP data for plotting. The CTDTu data presented only displays the downcast. Shown in Figure 2Ch3, water sample total suspended material (organic and inorganic) is correlated against CTDTu turbidity with an r-squared of 0.9 to produce suspended concentrations values for the entire CTDTu profile with depth. Technically, SSC values shown by the CTDTu are the total suspended organic and inorganic material, but for simplicity, they will be referred to as suspended sediment concentration (SSC). The difference between CTDTu SSC and water sample organic and inorganic material are also present in the results. More samples were available for the total suspended material than the LOI organic and inorganic due to the inclusion of the Millipore filters. CTDTu SSC values are useful compared to the water samples as they cover a full profile of the water column with depth while water samples only capture discreet points in the water column (maximum three) with depth.

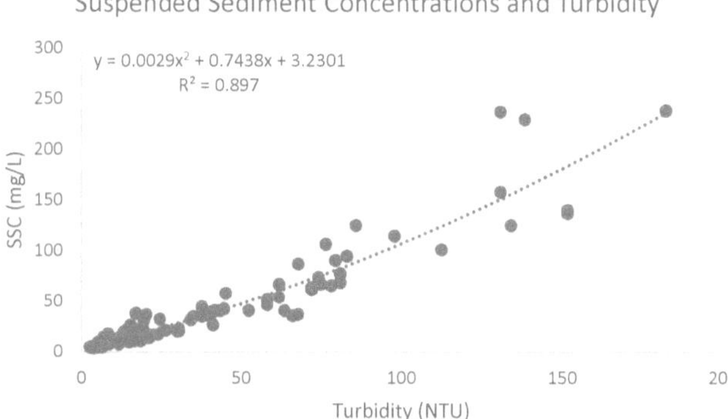

33 Figure 2Ch3: Correlation of water sample SSC and CTDTu turbidity from August 2019 and May 2019 Field Campaigns.
This equation was used to estimate the SSC for each turbidity profile taken with the CTDTu displayed later in this report.

Methods for Analyzing ADCP Flow

To display the water column velocity magnitudes and flow directions of ADCP transects, first raw ADCP data must be processed. ADCP data were collected in WINRiver II, exported, and processed into velocity magnitude and direction in MATLAB using scripts adapted from Colin Rennie (Rennie et al., 2002; Rennie and Rainville, 2006; Aberle et al., 2017; Moradi et al., 2019) as well as literature from Dinehart and Burau (2005). The ping sampling rate was every 45 millisecond that categorized the raw data into 0.5 m depth bins. MATLAB processed data was compared against WINADCP software visualization; however, processing the data in MATLAB allowed for greater processing quality control and flexibility. The following paragraph provides a condensed overview of the MATLAB ADCP processing. Further details, theory, and scripts are also available upon request.

The ADCP emits an acoustic signal into the water and measures the Doppler shifted frequency of the returning signal that has bounced off scatterers (suspended material) in the water column (Gordon, 1996; Greenwood, 1984). Four different beams in Janus configuration measure three directional components (x, y, and z) and the error relative to the instrument. The ADCP also records the echo intensity of the signal emitted from the scatters as a biproduct (Gordon, 1996). In MATLAB, raw (x, y, and z) velocity vectors are adjusted to true north by applying the magnetic declination. Velocity values 6% off the bottom are disregarded due to sidelobe interference when acoustic beams are positioned at 20 degrees (Yorke and Oberg, 2002). The velocity of the vessel derived from the bottom track or the external GPS must be removed from the water column velocities to depict the velocity within the water column (Rennie et al., 2002; Dinehart and Burau, 2005; Wagner and Mueller, 2011). After an initial comparison of the bottom track and GPS corrections, the external GPS was chosen to correct for vessel motion for the Skeena data. In the case of high bedload transport velocities at the bed, the external GPS may provide a more accurate measure of vessel movement (Wagner and Mueller, 2011). For the Skeena Estuary, there may be substantial movement at the bed that could add to the bottom track velocity in addition to vessel motion and at times the bottom track dataset had large gaps due to choppy waves. Therefore, the external GPS was used to remove vessel motion from the velocity components. A moving average every ten seconds was applied to the GPS vessel east and north velocities to smooth out unwanted outliers. After missing values were interpolated in the north, east, and upward velocity components, the velocity magnitude was computed. Next, the atan2 function calculates the direction of flow using the east and north velocity components. This produced a cross-channel matrix of velocity magnitude in ms^{-1} and flow direction in degrees every second into 0.5 m depth bins starting below the blanking distance and depth of instrument deployment (~1.7m) to 6% above the channel bottom.

Flow directions are visually analyzed over a tidal cycle transect to assess flow bidirectionality. Depending on the passage orientation, the velocity vector representing inflow and outflow (North for T and east velocity vectors for SDE) are used to compute mean inflow and outflow velocity at the surface, bottom, and different sides of the channel. The mean bottom was calculated by computing an average of all velocities below the channel mean depth during a given transect. Capturing slightly different portions of the ebb and flood tide during different surveys would introduce bias in the final mean or comparison of ebb vs flood strength. Long term surveys (over multiple days and tidal cycles) would be necessary to produce a mean inflow or outflow velocity value more accurately. Therefore, velocity values can be used to discuss local differences in the flow within a given transect but were not used to compute a mean tidal value of the surface or bottom within a given location. At any point when local tidal velocities are computed (for select locations of T and SDE in the results), information regarding the specific point of the tidal cycle is available in the appendix.

ADCP Sediment Inversion

To display suspended sediment flow within the water column, ADCP echo intensity can be converted to sediment concentration (SSC) when corrected for attenuation and other factors within the water column through a sediment inversion process (Bradley et al., 2013; Topping et al., 2007; Venditti el al., 2016 and Gartner, 2004). The theoretical equation synthesizes all the factors that influence backscatter signal loss and attenuation that must be considered when calculating SSC. This includes water temperature, salinity, distance of the signal from the transducer, turbidity (sediment density and concentration), and grain size type (Venditti et al., 2016 and Thorne and Hanes, 2002). MATLAB scripts for applying the sediment inversion were written by Sophie Hage, Maria Azpiroz-Zabala, Age Vellinga, and Stephen Simmons at the University of Hull for the Squamish estuary in southern British Columbia using an iterative process. The scripts were applied with Skeena estuary specific inputs derived from the ADCP, CTDTu, water samples, and literature for a given location within the estuary over the tidal cycle. A detailed description of the methods used to develop the sediment inversion scripts can be found in Hage (2019) and Simmons et al. (2020). The paragraph below provides a brief description of the inversion and Skeena specific inputs.

Echo intensity must first convert into linear backscatter counts (Simmon et al., 2020). This conversion was made using echo intensity from beam one of the Skeena ADCP data as well as using a Teledyne instrument specific transducer constant, and an instrument specific noise level recorded in Deines (1999). Next, the range from the transducer and near-field correction is applied using the equation presented by Downing et al. (1995). The near-field correction corrects for the spherical spreading of the signal whereby the signal level increases approaching transducer except for in the final meter (Downing et al., 1995). The near-field correction requires values such as instrument depth, beam angle, and bin size recorded as metadata with the ADCP during fieldwork. For the near-field calculation, the speed of sound in water was taken from the CTDTu, the necessary transceiver radius is recorded in Downing et al. (1995), and the specific frequency of the ADCP can be found in the RDI manual. The backscatter properties of sediment for the inversion are computed based on the D50 and SD using the equation developed for a mixture of sediment types by Moate & Thorne (2012) and Thorne and Hanes (2002) and used to create a normalized frequency distribution of grain sizes for the given grain size diameter through a log-normal distribution (Hage, 2019). The sediment density input is from Moate &

Thorne (2012). The computation of the d50, min, and maximum grain size for the Skeena water column is described in further detail in the following paragraphs. A calibration instrument specific constant is applied next using instrument specific data described in (Deines, 1999), or ADCP metadata during field collection. Finally, attenuation coefficients can be applied using the equation from (Francois and Garrison, 1982) for water and from (Urick, 1948) for sediment attenuation. The computation of water attenuation uses mean water column temperature, density, depth, and salinity calculated from the transect mean of CTDTu profiles. Ph is derived from the average of PH in estuarine waters described by Millero (1986). An iterative process is used to apply sediment attenuation and compute the final ADCP backscatter derived sediment concentration (Hage, 2019). A final constant was applied to bring inversion values into the appropriate mgL^{-1} order of magnitude for the Skeena and comparable to the CTDTu values.

To calculate the sediment attenuation coefficient, the d50 grain size of the water column is needed (Simmons et al., 2020). Suspended sediment grain size samples were retrieved for the Skeena Estuary at the SDE in May, producing a median d50 of 6.9 μm of very fine silt according to the Wentworth scale. May SDE suspended sediment grain sizes ranged from clay to coarse silt. These samples do not include the area directly approaching the bed or the bedload as the water sampler was positioned above a weight and CTDTu causing the deepest depths water sampled to be ~2 m above the bed. Also, river inflows were low during May sampling, so grain sizes may often be higher than those measured. The ADCP SSC results from the inversion based on a d50 of very fine silt will likely over-predict the SSC approaching the bed as the grain size increases. This is because the backscatter is receiving a stronger signal from the larger grain size clast opposed to a higher concentration of fine sediment. This was evident in the sensitivity analysis of different d50 values used in the inversion. The 6.9 μm d50 is also smaller than even the smallest values used for the d50 inversion in Hage (2019) for the Squamish. Thus, to improve the ADCP inversion in SDE, a rough estimate of the bedload component was needed. Grain size samples at the bed near the SDE location are sand dominant (98-100% sand). Further seaward in the estuary surface samples often have a proportion of very fine sand and occasionally fine sand indicating that some of this material is in transport seaward or landward. Peak ebb and flood velocities were over 1 and under 1.2 ms^{-1} which, along the Hjulstrom curve (1935), transport up to the very fine sand and fine sand boundary (~100 to 125 μm). Thus, according to the Hjulstrom curve (1935), at the above velocities, everything larger than very fine sand would fall out of transport, leaving the seabed sand dominant at the SDE location. Meanwhile, the largest sediment size recorded in the SDE suspended load above the bed is 55 µm, which is coarse silt. Thus, we can infer that the bedload likely composes of greater than 55 µm coarse silt to very fine sand less than ~125 µm. An estimated mean of 62 µm at the coarse silt and very fine sand boundary was chosen for the bedload component. Based on 1981-2010 modelling results, the annual proportion of suspended load to bedload is 56/44% respectively. Based on this ratio, the suspended load d50, and the estimated mean bedload grain size, a mean grain size for the water column of 31 µm was applied to the ADCP sediment inversion in SDE. Further seaward within the estuary, finer sediments dominant the water column. Therefore, after a brief sensitivity analysis, the measured suspended sediment d50 of 6.9 µm was applied to T survey transects. The minimum grain sizes for SDE and T based on the suspended sediment grain size distribution is 1 µm, and the maximum was estimated at 250 µm in case any fine sand mobilizes as bedload under higher flows.

The sediment inversion method assumes that there is a uniform grain size distribution throughout the water column (Hage, 2019; Sassi et al., 2012). Effects on backscatter values due to particle concentration cannot be separated from those due to changes in particle size using a mono-frequency ADCP (Sassi et al., 2012). Changes in grain size within the water column can appear similar to changes in concentration within the acoustic signal of a mono-frequency ADCP. A fining of sediment can result in increased sound attenuation due to viscous loss (Topping et al., 2007; Wright et al., 2010). Meanwhile, increased grain size clasts may increase scattering and increase the strength in the backscatter of sound (Topping et al., 2007; Wright et al., 2010). Thus, a larger d50 used in the inversion would produce a stronger sediment concentration using the ADCP backscatter and inversion method. The Skeena estuary is highly dynamic, and grain size evidence suggests that grain sizes vary throughout the water column. Thus, the assumption of a uniform grain size distribution would introduce error in the ADCP SSC results. The mean grain size rather than the suspended d50 would likely cause a decrease the appearance of concentration in the upper water column where there are finer sediments than the d50 and increase the bottom where grain sizes may be higher. Under higher river inflows in August, grain sizes would likely increase throughout the water column however no suspended sediment grain size samples were collected in August to apply to the sediment inversion. Further studies with suspended sediment grain size sampling at different locations in the estuary would be necessary to improve the accuracy of the sediment inversion.

The sediment inversion visually shows sediment movement over the tidal cycle as well as areas of high and low sediment suspension throughout the water column that has been corrected for attenuation and the near-field factor opposed to raw backscatter values. Only T and SDE with consistent survey transects in both seasons over the tidal cycle will present suspended sediment flow based on the sediment inversion. The CTDTu will be used to refer to actual SSC concentrations for a given area. Due to uncertainty an inherent assumption within in the ADCP technique that may cause visualizations to differ from reality, ADCP SSC sediment inversion will be presented with overlain CTDTu casts.

Suspended Sediment Load (Qs) Calculation

One of the debated questions regarding the Skeena estuary is the diversion of suspended sediment flow into each estuarine passage. To address this question, tidally averaged mean suspended sediment load (Qs) was calculated for each passage using tidally averaged CTDTu SSC and the mean volume derived from hydraulic geometry. Refer to Table 2Ch3Appx within the appendix for hydraulic geometry calculations and to the results for final mean SSC and Qs. Hydraulic geometry calculates mean passage volume based on a multiplication of mean width, depth, and velocity. Velocity is the largest controlling factor on the fluctuations in volume over the tidal cycle within a passage. Additionally, mean tidal velocity may slightly differ between passages, but this difference is challenging to confirm using the current data. Therefore, due to varied sampling coverage over the tidal cycle [29], one mean velocity based on sampling over a full tidal cycle in a landward and seaward passage was used to estimate mean Qs for all passages. This is to isolate mean conditions and remove any variance in sampling inconsistencies such as capturing different portions of the tidal cycle between passages. More specifically, the velocity from SDE and T (with the most consistent coverage over the entire tidal cycle) was averaged over all available seasons and points in the tidal cycle and applied to all passages as the mean velocity. Tidal and seasonal mean depth was calculated from ADCP data for each passage.

---

[29] Some passage sampling terminated prior to capturing higher flow velocities. For example, M had an ADCP malfunction during an ebb transect leaving only CTD SSC data and no velocity magnitude values available for the ebb tide. August T surveying began slightly later in the tidal cycle than SDE producing lower ebb velocities.

Transects in May and August were surveyed as close as possible to eliminate location bias. Holding depth constant between seasons also helped to emphasize changes in Qs due to SSC rather than capturing a slightly deeper transect in a different season. Mean channel width for each passage was calculated using google earth imagery. CTDTu SSC averaged over the entire cross-channel, and available tidal cycle produces SSC values for each location and season. Finally, the tidally averaged discharge and SSC multiply to produce the mean channel Qs.

Calculation of Stratification Parameters

CTDTu salinity casts were visually analyzed over a tidal cycle within multiple passages and used to discuss estuarine stratification. Stratification was also indicated through the calculation of a stratification parameter in landward (SDE) and seaward (T) areas of the estuary with the most consistent data coverage over the full tidal cycle in May and August. A vertical stratification parameter (VSp) is calculated from the difference in salinity in the top and bottom of the water column divided by the average throughout the water column (Valle-Levinson, 2011; Fontes et al., 2014; Hansen and Rattray, 1966). More specifically for the Skeena data, the water column average salinity is derived from all CTDTu casts taken across the channel for a given ebb, slack, and flood transect with CTDTu data. The difference in salinity (ΔS) was derived from an average of the first 2 m of all CTDTu casts taken across the channel under a given tidal stage subtracted from the average conditions below the mean channel bottom. VSp is more sensitive to changes in salinity within fresher waters and only compares the surface and bottom without consideration of the abrupt or gradual approach taken in between. For the vertical stratification parameter, 0 indicates well mixed and 1 and over indicates high stratification within estuaries (Valle-Levinson, 2011). For this research and based on Valle-Levinson (2011) & Hansen and Rattray (1966), VSp values under 0.1 will be considered well mixed, over 1 is highly stratified, and 0.1-1 is weakly to appreciably stratified.

Typically, stratification parameters described by Valle-Levinson (2011) are a net, tidally averaged, metric of vertical stratification. However, in theory, data is available for a stratification parameter to be applied to local ebb, slack, and flood portions of the tidal cycle vertically as well as horizontally. Gradient Richardson's numbers [30]were calculated for CTDTu profiles during the ebb and flood within August Telegraph Passage and compared against local VSp values to test that the stratification parameter method of Valle-Levinson (2011) was accurate in describing the stratification during local portions of the Skeena tidal cycle. Within nearly all CTDTu profiles analyzed, stratification occurred within the upper 10 m of the water column with the bottom of the water column being more saline and well mixed. Gradient Richardson numbers indicate a higher stratification with the majority (over 50%) of the water column in the upper 10 m displaying stratification in line with higher stratification indicated by the VSp parameter. More specifically, during the August T ebb, a VSp of 0.7 indicate weakly stratified conditions and gradient Richardson numbers are above 0.25 (indicating stratification) within 35% of the water column between 1.6-10 m depths. During the August T flood tide, a VSp of 1.1 indicated highly stratified conditions and Richardson numbers are over 0.25 within 53% of the upper 10 m of the water column. Gradient Richardson was only used for an initial validation of the stratification parameter and will not be within the results. Instead, all VSp values are presented with associated CTDTu profiles within the results.

Although no such parameter was described by Valle-Levinson (2011), using a similar approach, the horizontal stratification parameter (HSp) is calculated from the difference in salinity within the first 2 m of the surface of salinity casts on one side of the channel compared to the other side, divided by the total channel mean surface salinity. Different sides of the channel often had different depths, and a deeper depth usually produced a more saline mean salinity. Only the surface of the water column was used to calculate HSp in order to isolate horizontal

---

[30] According to Winant and Velasco (2003) and Prandle (2009), Ri measures the local buoyancy relative to the shear and can be measured as $Ri = \frac{g\left(\frac{\partial p}{\partial z}\right)}{p\left(\frac{\partial u}{\partial z}\right)^2}$. whereby g is the acceleration of gravity, p is mean density, ∂z is the change in depth, ∂p is the change in density, and ∂u is the change in velocity. CTD density was averaged into the same 0.5 size depth bins as the ADCP velocity to compute the Richardson number between each ADCP depth bin in the location where the deepest CTD casts were consistently taken within a given transect location. 0.25 is the critical value for Richardson numbers that differentiate between vertical mixing and stratification (Prandle, 2009). A Richardson number under 0.25 indicates that turbulence overcomes density layering allowing for vertical mixing to occur (Prandle, 2009). Richardson's numbers can vary significantly over the tidal cycle, riverine inputs, and location.

stratification due to flow on either side of the channel and not depth. Therefore, HSp could only be calculate in transects where profiles were taken on opposing sides of the channel.

ADCP Discharge, Tidal Prism, and Estuarine Flushing

Comparing the tidal prism to riverine inflow is another method to assess an estuary's stratification on a larger scale. When the tidal prism is more than an order of magnitude higher than the riverine inputs over the tidal cycle, the area is likely vertically homogenous, well mixed, or weakly stratified (Valle-Levinson, 2011). When the order of magnitude of freshwater and tidal prism volumes match, estuaries are typically partially mixed. Finally, when the freshwater inflow volume is more than an order of magnitude than the tidal prism, the estuary is usually highly stratified (Valle-Levinson, 2011). For the Skeena, the tidal prism is calculated from the change in morphology in the braided river to become tidally drowned to the Skeena River directly downstream of Ecstall (Figure 1Ch3). The tidal prism covers essentially the same area as the TDSR sub-environment from chapter 2. Tidal prism and tidal flushing calculations assume an area that is well mixed (Lakhan, 2003). Therefore, this tidal prism calculation does not include the intermediate and outer estuarine passages that showed stratification as will be described later in this chapter (Figure 1Ch3). According to conventional methods of calculating the tidal prism through the basin surface area multiplied by the tidal range (Valle-Levinson, 2011). With channel discharge, velocity, salinity, width, and depth available at two locations near or within the tidal prism extent, further calculations on tidal flushing can be conducted. The exchange ratio is a measure of the intertidal volume at a given location divided by the intertidal and low tide volume (Ketchum, 1951). Escaping volume is a measure of the riverine inflow over a tidal cycle divided by the percent freshwater and should increase seaward (Ketchum, 1951). Both exchange ratio and escaping volume have been calculated to compare SUE and SDE locations within and just outside of the tidal prism area.

For the calculation of the tidal prism and estuarine flushing over a tidal cycle, the discharge is calculated at peak flood, ebb, and slack points along the tidal cycle at the SUE and SDE locations. MATLAB scripts were established to calculate channel discharge based primarily on the literature of Simpson (2001). In order to calculate channel discharge, the distance to the bank, surface, and bottom that were not captured by the ADCP must be extrapolated while the rest of the discharge can be calculated from the summation of the ADCP cell depth, width, and velocity product. The distance to the banks from the end of the ADCP profile was measured in google maps. For each bank, the bottom depth is multiplied by the last ADCP ensemble average velocity and the distance to the bank. This is all multiplied by 0.707, to account for an assumed reduction of velocity due to increased drag and friction as the water depth decreases and interacts more with the bed (Simpson, 2001). Finally, this value is divided into two under the assumption that the depth at the bed decreases linearly to the bank producing a triangular discharge area (Simpson, 2001). Next, the discharge for the rest of the channel that was covered by the ADCP with extrapolated top and bottom water column areas must be computed and added to the discharge calculated for the channel sides. The velocity magnitude and bottom depth values are averaged into minute-long intervals to produce water profiles with less random noise that are easier to fit a curve for extrapolation of the top and bottom of the water column. Chen et al. (2016) highlighted how the commonly used power-law method did not accurately capture the water profile trends in estuarine environments. They recommended trying different methods and choosing an extrapolation method that was the best able to capture the water column statistically (Chen et al., 2016). After multiple experiments, eventually, a constant method for the surface and a polynomial extrapolation for the bottom of the water column was applied for the Skeena data. The constant method is where the first and last recorded values of the ADCP are set as the missing velocities at the surface and bottom (Yorke and Oberg, 2002). For each depth cell in the matrix of each one-minute averaged ensemble, the velocity (ms^{-1}) is multiplied by the distance the boat travelled between ensembles (cell width in meters) and the depth cell size (m). This computes a matrix of discharge measurements in ($m^3 s^{-1}$). Each depth and ensemble cell in the matrix sum into one discharge amount that is added to the missing side area discharges.

Estuarine Classification Methods and Calculations

Geyer (2010) used a dimensionless tidal and river unit to create a prognostic estuarine classification for many different estuaries based on stratification. The dimensionless tidal unit (Td) is calculated from tidal current amplitude (Ut) divided by a densimetric velocity scale. Since the tidal amplitude is comparable between August and May transects (ECCC, 2020), the highest flood or ebb mean cross-sectional velocity recorded at a given location in the estuary over both seasons minus the lowest slack velocity was used to produce the tidal current amplitude (Ut). This approach was chosen to try to avoid biases from capturing slightly different points and strengths in the tidal cycle between May and August. The dimensionless river unit (R) is calculated from the integral mean fluvial velocity (Uf) divided by a densimetric velocity scale. Estuary cross-sectional area (*Acs*) is calculated based on the mean depth for the estuarine location multiplied by the channel width measured in Google Earth. River discharge (*Q)/Acs* is used to calculate *Uf.* River discharge (*Q*) is calculated using the mean modelled HydroTrend discharge output for the Skeena watershed of the same month and within $100\ m^3 s^{-1}$ (for subbasin 2) of the discharge measured at Usk hydrometric station on the day of the ADCP survey. The discharge measured at Usk station was not used as the value to represent the river inflow as it excludes a substantial coastal portion of the watershed. Refer to chapter 1 for further details on modelling discharge and other parameters for the Skeena watershed. The densimetric velocity scale is (β g So h)^(1/2) whereby β is the coefficient of saline contraction n (0.00077), g is the acceleration due to gravity (9.8 $ms^{-1}s^{-1}$), h is the water depth, and So is the mean salinity for a given cross section within the estuary. Geyer (2010) describes an estuary as highly stratified when R is over 0.1 regardless of Td. However, a ratio of R and Td is used to differentiate between well-mixed and highly stratified estuaries. Skeena Td and R result are applied to the Geyer (2010) prognostic estuarine classification in the discussion.

Results

Suspended Load Divergence through the Estuary

Tidally averaged suspended sediment load (Qs) for the SDE, IV, and T sites are under moderate flow conditions and low flow conditions for SDE, IV, T, and M (Table 1Ch3; Figure 3Ch3). The SDE estimate yields the total sediment load delivered to the estuary by the Skeena River, including its last tributary Ecstall. IV, T, and M estimates yield the sediment load through each of the three passages and thus can be used to examine sediment load divergence in the estuary (Figure 1Ch3). Table 1Ch3 shows the tidal CTDTu mean SSC for each channel as well as the passage volume necessary to calculate loads. Refer to Table 2Ch3Appx in the appendix for the hydraulic geometry variables used to calculate passage volume. Shown in Figure 3Ch3, sediment load at SDE is 500 *kg s*⁻¹ under low river inflows and 1150 kg s⁻¹ during moderate river inflows. The combined Qs for the three passages under low river inflows is 513 kg s⁻¹, with Marcus Passage contributing the most (255 *kg s⁻¹*, 50%), followed by Telegraph Passage (205 *kg s⁻¹*, 40%), and Inverness Passage (53 kg s⁻¹, 10%). Under moderate flow conditions, Qs through Telegraph doubles (430 *kg s⁻¹*), while Qs through Inverness stays nearly the same (55 *kg s⁻¹*). Based on the SDE estimate of Qs under moderate flows, a minimum estimate of Qs through

Marcus Passage would therefore be 665 $kg\ s^{-1}$ (58%), assuming all remaining material arriving in the estuary is transported seaward.

Table 1Ch3 also shows the breakdown of organic vs inorganic content of surface dominated water samples. Under higher river inflow August conditions, the inorganic concentration increase substantially more than the organic. Both inorganic and organic sample content is higher landward within the estuary. As inorganic content decreases in the seaward passages with a large volume, the organic concentration remains relatively consistent between all passages. CTDTu derived SSC is generally higher in narrower passages with a lower volume (e.g.: IV) and further landward prior to the divergence into multiple passages (e.g.: SDE). CTDTu profiles allowed for greater averaging over depth and are used over the water sample total values.

11 Table 1Ch3: Different estuarine passage tidal mean organic and inorganic water sample concentration, CTDTu profile total suspended concentration, passage volume, and suspended load.

Date	Location	WS Organic Concentration Mean (mg/L)	WS Inorganic Concentration Mean (mg/L)	CTD Total Concetration to Depth (SSC) (mg/L)	Mean Passage Volume Q (m3/s)	Tidally Averaged Suspended Load Qs (kg/s)
5/2/2019	Skeena Downstream Ecstall (SDE)	2.79	14.93	18.51	26997	500
5/2/2019	*Inverness Passage (IV)*	*2.81*	*14.06*	*16.70*	*3199*	53
5/3/2019	Telegraph Passage (TP)	2.36	5.39	10.15	25549	205
5/1/2019	Marcus Passage (M)	2.00	2.81	17.40	25117	255
8/20/2019	Skeena Downstream Ecstall (SDE)	6.33	51.33	42.60	26997	1150
8/19/2019	Inverness Passage (IV)	4.41	17.57	17.24	3199	55
8/22/2019	Telegraph Passage (TP)	4.35	9.32	16.68	25549	426

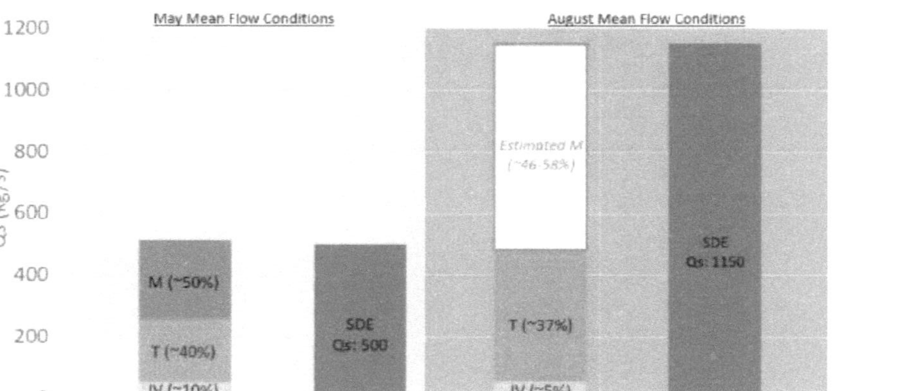

34 Figure 3Ch3: May (left) and August (right) sediment load (Qs) divergence into different passages (M (grey), T (orange), &IV (yellow)) compared to SDE (blue).

Flood and Ebb Salinity Profiles throughout the Estuary

Refer to Figure 4Ch3 to see the ebb and flood salinity profiles for four passages within the Skeena Estuary. Under August flows, salinity is highest at the seaward location IV and T (10-30 PSU), moderate in SDE (2-12) and lowest at location SUE (0.3-0.4) closest to the Skeena River. Stratification, visually observed by two vertical regions of differing salinity with a sharp increase in intermediate waters, can be observed at the SDE, IV, and T locations whereas SUE is vertically homogenous. Stratification occurs abruptly during flood tide and more gradually during ebb tide. The difference in salinity between the top and bottom of the water column increases seaward from the river. The difference in salinity is the most significant during the flood tide (Figure 4Ch3).

Under low May river inflows, salinity in the upper water column is higher than under August flows at all locations. At the most seaward locations IV, T, and M salinity ranges between 21-30 PSU and between 11-20 PSU at SDE. A profile for SUE under lower May flows is not available. All salinity profiles show a slight increase (by about 2-6 PSU) in salinity with depth but are generally well mixed. However, due to this increase in salinity with depth, SDE, IV, M, and T are not considered vertically homogenous. The largest salinity increases in depth occur at T and SDE during the ebb tide.

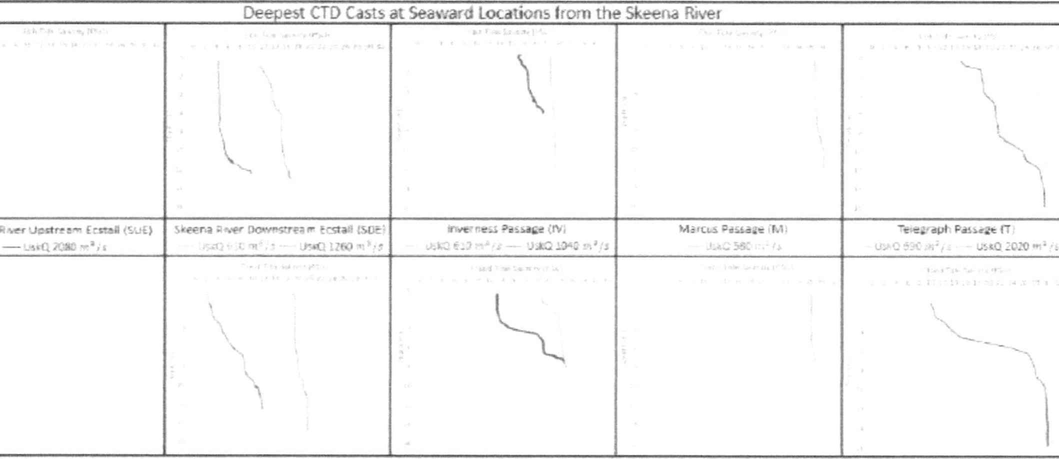

35 Figure 4Ch3: Figure 4Ch3 displays flood and ebb salinity profiles with depth under May (low river inflows around high tide) and August (moderate river inflows around low tide) flow conditions at various locations across the estuary.
Light blue represents May, and dark blue represents August profiles. USKQ stands for the discharge at Usk during the day the CTDTu casts were taken at a given location within the estuary. The deepest CTDTu cast across the channel was select. If two profiles were near equivalent in depth, the one closest to the channel center was selected. Data are presented with ebb tide profiles in top graphs and flood tide below. The furthest landward location is presented to the left. Profiles are not available at all estuary locations over multiple seasons. For example, due to high winds, Marcus was not surveyed under moderate river flows. The furthest landward location, SUE, was not surveyed during flood tide.

Tidal Transition Dynamics throughout the estuary

ADCP cross-sectional velocity fields display the flow directions within each passage during a tidal transition and around slack water (Figures 5Ch3 and 6Ch3). During peak ebb and flood velocities, flow is unidirectional in the tidal flow direction, except for a few eddies and countercurrents directly behind bedrock outcrops in SDE. Bidirectional flow observed surrounding slack tide in all surveyed channels (Figures 5Ch3 and 6Ch3), highlight complex tidal flow dynamics that differ between river inflows and locations within the estuary. Generally, the contrast in the flow bidirectionality is starker seaward in the estuarine passages (e.g.: T) and more mixed landward (e.g.: SDE). The disproportionality in velocity magnitudes across the channel observed already during the flood and ebb period that leads to varied transitions to the following tide during slack. During slack tide, SUE, SDE, IV, M, and T transition to the next tidal flow direction last in area of the channel that had the highest velocity during the previous ebb or flood tide. The following sections focus on the slack tide transition in flow between different channels.

Under August slack conditions, at the furthest landward location SUE, the water column is vertically homogenous and the flow transitions via horizontal separation towards the flood tide (Figure 5Ch3A). While flow is still directed seaward towards the NW shore, flow is already flooding (landward) towards the shallower SE shore. Under similar river inflows (within 100 $m^3 s^{-1}$), further seaward at T, the transition towards flood occurs through vertical separation (Figure 5Ch3D). At IV, under substantially lower (~1000   lower USKQ inflow) river inflows, the water column also transitions vertically across the channel during portions of the slack tide (Figure 5Ch3C). Thus, seaward IV and T locations show a vertical flow transition with seaward flows at depth. Meanwhile, the more landward transect at SUE is transitioning horizontally. Finally, at SDE location between SUE and T in the estuary, flow transition occurs somewhat vertically and horizontally across the channel (Figure 5Ch3B). There is evidence of the incoming flood pushing at the bed on both the north and south side of the SDE channel, and indeed, the landward flood flow still dominates the north side of the channel. At the surface and all depths on the southern side of the channel, flow is still ebbing seaward. USKQ inflow for the SDE location is similar to IV and nearly half that of T and SUE. Thus, river inflow, as well as the greater distance from the incoming river, could be contributing to the vertically and horizontally mixed transition at SDE.

Under low river inflows, flow transitions occur via horizontal separation at all locations (Figure 6Ch3). Figure 6Ch3 depicts times when the tide is transitioning from flood to ebb. At locations SDE and T (Figures 6Ch3A and 6Ch3C), flow begins ebbing at the N and W shore, respectively, while flow remains flood directed in the deepest parts of the channel towards the S and E shore, respectively. The transition at M is markedly different (Figure 6Ch3B). Ebb flow occupies the channels towards the SE shore, while vertical separation occurs at the NW shore, separating the remaining flooding (landward) flow from southward directed flow in the bottom water column. To the south of M is another bedrock confined passage that appears to be influencing the slack tide flow. Thus, M could be described at tridirectional at slack tide with water parcels outflowing towards Arthur and Malacca Passage along the bottom, landward towards the Skeena River in the NW, and seaward towards Chatham Sound in the southeast.

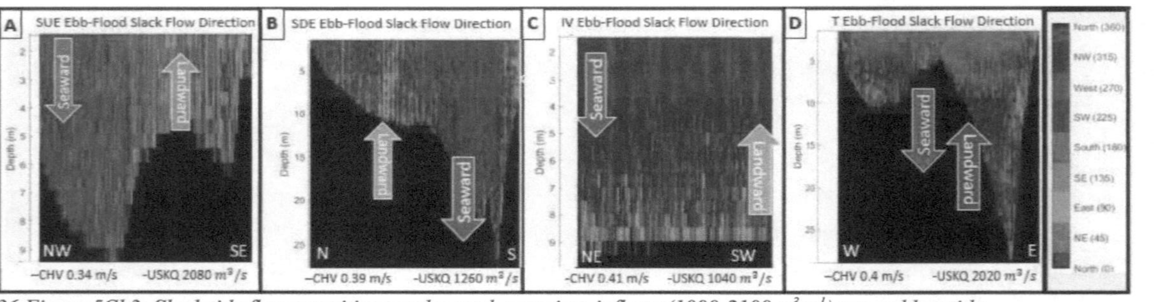

36 *Figure 5Ch3: Slack tide flow transitions under moderate river inflows (1000-2100 m^3 s^{-1}) around low tide.*
USKQ stands for Usk discharge during the day of sampling and CHV stands for the mean cross-channel velocity for the transect.
Arrows are colour coded to match the flow direction legend and point landward and seaward based on ADCP cross-channel transect orientation. Location A, presented first, is closest to the river and transect D, presented last, is furthest seaward.

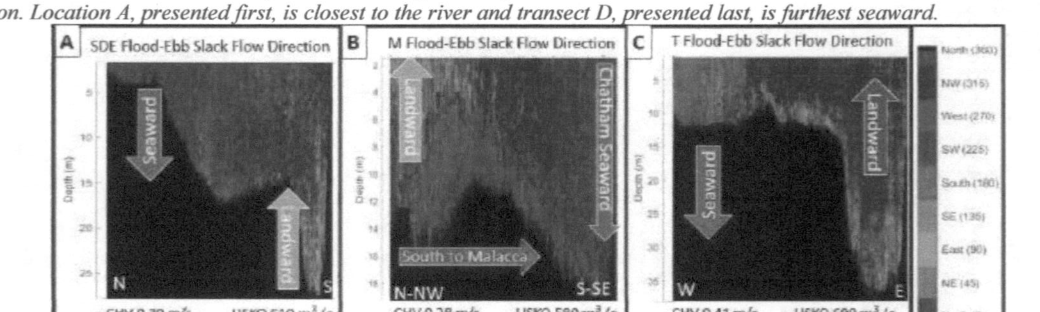

37 *Figure 6Ch3: Slack tide flow transitions under low river inflows (500-1000 m^3 s^{-1}) around high tide.*
Arrows are colour coded to match the flow direction legend and point landward and seaward based on ADCP cross-channel transect orientation. Location A, presented first, is closest to the river and transect C, presented last, is furthest seaward.

Detailed Tidal Transition Dynamics in a Seaward Passage (T)

August flow conditions (Figure 7Ch3A-F) depict vertically dominated and horizontally influenced stratification and flow during the ebb to flood low tide transition. Refer to Figure 1Ch3Appx in the appendix for the exact timing of flow figures during the tidal cycle. During decelerating ebb tide (Figure 7Ch3A), flow is fully unidirectional until ~3 hours before the lowest point of the surveyed slack tide when flooding occurs in the shallow, western segment of the channel close to the bottom (Figure 7Ch3B; Figure 1Ch3Appx). Flow velocities are fastest close to the surface above both the shallow and deeper areas of the transect (Table 2Ch3). Under these conditions, the horizontal stratification parameter (HSp) and vertical stratification parameter (VSp) indicate moderately stratified conditions (0.74 and 0.69, respectively (Table 2Ch3) vertically and across the channel. This translates to maximum salinity differences of 11 PSU between west and east segment and up to 14 PSU between the surface and bottom waters. During lowest velocity point of the surveyed slack tide, the entire channel is flooding at the bottom while it is still ebbing at the surface (Figure 7Ch3C-D). Flow is faster in the western segment than in the eastern (Table 2Ch3). The slack tide water column is highly stratified (VSp = 1.1) resulting from a decrease in salinity at the surface compared to the ebb and the formation of a marked halocline (Figure 4Ch3). Horizontal stratification is negligible (HSp = 0.1). Following slack tide under accelerating flood conditions (Figure 7Ch3E; Figure 1Ch3Appx), flow is fully unidirectional after only ~1.5 hours after slack tide with strongest velocities in the deeper eastern segment. Notice that flow bidirectionality prevails longer during the decelerating tide. During the accelerating flood, velocities in the surface layers remain lower than at depth (Table 2Ch3). Vertical stratification remains high (VSp=1.1) with a salinity difference of up to 20.3 PSU between the upper and lower water column. Horizontal stratification remains absent (HSp = 0.1).

In May, flow transitions via horizontal separation from flood to ebb tide (Figure 7Ch3G-L), and the water column is well mixed throughout. During the flood tide (Figure 7Ch3G-H), flow is uni-directional with low stratification numbers (VSp =0.16 and HSp= 0.03) and highest velocities in the deeper, eastern segment, close to the surface (Table 2Ch3). During slack tide (Figure 7Ch3J), flow transitions horizontally across the channel, with ebbing starting in the shallower western segment. This causes a slight increase in HSp (0.14), though remaining in the well-mixed classification. The observed salinity differences are up to 3.8 PSU, only between east and west segment. During the ebb flow (Figure 7Ch3K-L), when the flow is unidirectional (Figure 7Ch3G), HSp and VSp are low (0.06 and 0.07, respectively) indicated well-mixed conditions. Velocities remain highest above the shallower, western segment (Table 2Ch3). In May, bidirectional flow begins to propagate and lingers ~ 1hr before and after the lowest cross-channel velocity low tide (Figure 7Ch3I-K, Figure 1Ch3Appx).

12 Table 2Ch3: T select CTDTu cast ebb, slack, and flood, salinity, velocity, VSp, and HSp. Dimensionless tidal and river unit will be addressed in the discussion along with figures from Geyer (2010). VSp calculations are described in the methods and based on Valle-Levinson (2010). Velocity and salinity were chosen from available ADCP transects with CTDTu casts that showed ebb dominant, flood dominant, and bidirectional slack flow (Figure 7Ch1B, D, F, H, J, L). The exact point of the tidal cycle is shown in Figure 1Ch3Appx. Velocities varied drastically with time lag from slack and by capturing different portions of the tidal cycle during a transect, therefore, mean tidal velocities were left as not available (NA). Velocities are taken from North-South velocity vectors for T based on channel orientation relative to the incoming and outgoing flow. As described in the methods, Sp values 0.1 and under are well mixed, over 0.1 but under 1 are weakly stratified, and over 1 are highly stratified. The tidal mean surface is fresher and faster flowing at the surface, especially on the western side of the channel.

T Variables	Units	August Ebb (B)	August Slack (D)	August Flood (F)	Tidal Mean	May Flood (H)	May Slack (J)	May Ebb (L)	Tidal Mean
HydroTrend Q (m3/s)	Q (m3/s)				2726.00				1197.00
Cross-Channel Mean Salinity	S_0 (PSU)	20.35	17.82	17.83	18.67	28.63	29.64	27.87	29.64
Bottom Mean Salinity	S_b (PSU)	29.50	30.18	30.58	30.08	28.25	30.42	28.67	30.42
Surface Mean Salinity	S_s (PSU)	15.44	10.42	10.26	12.04	26.23	27.26	24.34	27.26
West Surface Mean Salinity	S_w (PSU)	5.56	10.66	10.60	8.94	25.38	25.61	23.54	25.61
East Surface Mean Salinity	S_e (PSU)	16.98	9.88	9.88	12.24	27.08	29.40	24.17	29.40
Horizontal Stratification Parameter	HSp (-)	0.7	0.1	0.1	0.2	0.06	0.14	0.03	0.1
Vertical Stratification Parameter	VSp (-)	0.7	1.1	1.1	1.0	0.07	0.11	0.16	0.1
Cross-Channel Mean Velocity (North-South)	Ux (m/s)	-0.50	0.13	0.50	NA	0.6661	0.09	-0.54	NA
Bottom Mean Velocity	Ub (m/s)	-0.46	0.50	0.59	NA	0.7849	0.21	-0.90	NA
Surface Mean Velocity	Us (m/s)	-1.01	-0.28	0.20	NA	0.6974	0.13	-0.69	NA
West Surface Mean Velocity	Uw (m/s)	-1.37	-0.19	0.31	NA	0.1779	-0.23	-1.25	NA
East Surface Mean Velocity	Ue (m/s)	-0.86	-0.38	0.08	NA	1.0875	0.36	-1.66	NA
Tidal Current Amplitude	U_T (m/s)				0.75				0.75
Fluvial Velocity (Q/Cross-sectional Area)	U_R (m/s)				0.033				0.015
Dimensionless Tidal Unit $U_T/U_R\sqrt{AT}$	Td (-)				0.50				0.40
Dimensionless River Unit $U_R/U_R\sqrt{AT}$	R (-)				0.022				0.008

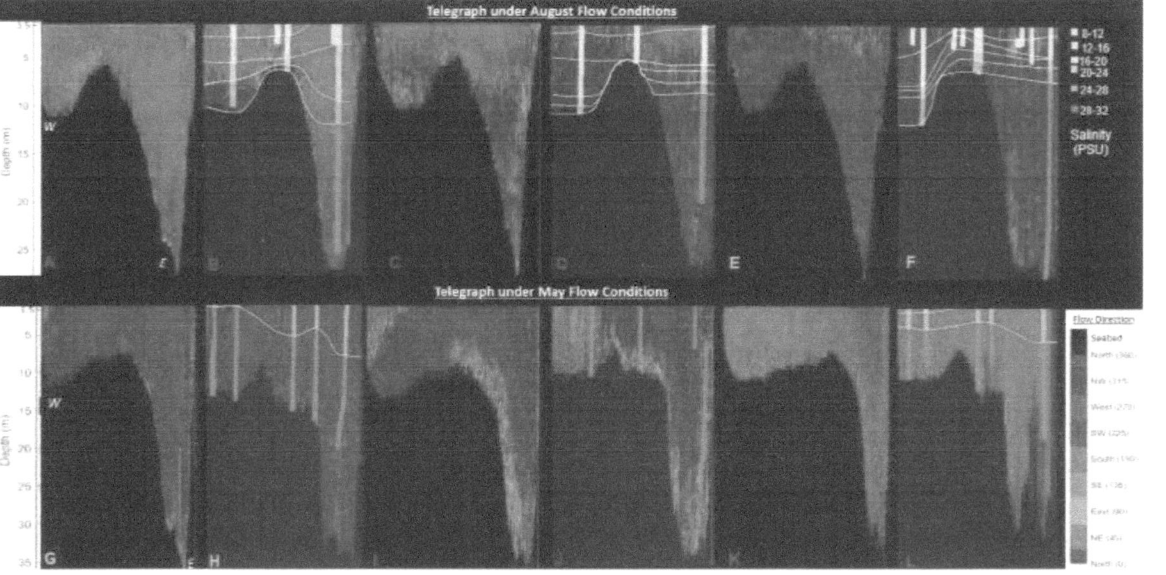

38 Figure 7Ch3: Telegraph flow direction with salinity profiles to indicate stratification.
White lines indicate salinity contours with the exact values and locations of the CTDTu casts presented over the ADCP transect flow direction values. August data from ebb to flood surrounding low tide under $2020\,m^3s^{-1}$ USKQ are presented in the top figures, and May flood to ebb surrounding high tide under $690\,m^3s^{-1}$ USKQ data are presented on the bottom. Refer to Figure 1Ch3Appx within the appendix for the exact point in the tidal cycle of A-L transects. A-L are highlighted in purple during the ebb, green during the slack, and yellow during the ebb. The transects were surveyed from the western to eastern bank with southern flows representing seaward (ebb) tidal outflow and northern flows representing landward (flood) tidal inflows.

135

Figure 8Ch3 displays suspended sediment concentration (SSC profiles (ADCP) and SSC casts (CTDTu)) for the T May and August transects shown in Figure 7Ch3. The overlain SSC casts can be used to assess the accuracy of the ADCP backscatter inversion visually. Generally, the ADCP SSC profiles agree with the CTDTu casts except for distinct horizontal bands of high SSC that show up in the ADCP data in intermediate waters in August, but not in the casts (Figure 8Ch3 D-F). This discrepancy within the sediment inversion ADCP may be due to several factors such as fish, turbulence, or changes in water density where these bands occur that are not related to water column sediment.

Generally, based on CTDTu casts and ADCP profiles, suspended material appears to be mobilized at the bed in May during higher velocity flows and settles from the surface during August flows. More specifically, in August flow conditions, higher SSC (15-30 mgL^{-1}) occurs in the fresher water to the west during decelerating ebb flow (Figure 8Ch3 B) and at the surface during low water slack tide and accelerating flood (Figure 8Ch3 D & F). This is more evident in the CTDTu casts but is also evident in the ADCP within Figure 8Ch3 B. Deep and more saline water ranges between 5-15 mgL^{-1}. In May during lower river discharge and well mixed conditions, SSC is low (5-15 mgL^{-1}) throughout the water column during the decelerating flood, high water slack tide, and accelerating ebb flow (Figure 8Ch3G-L). High SSC occurs close to the bed due to resuspension under accelerating ebb flows (Figure 8Ch3L), and higher velocity flood flows (Figure 8Ch3G). The same mobilization at the bed does not appear to occur during the higher flood and ebb flows in August flow conditions.

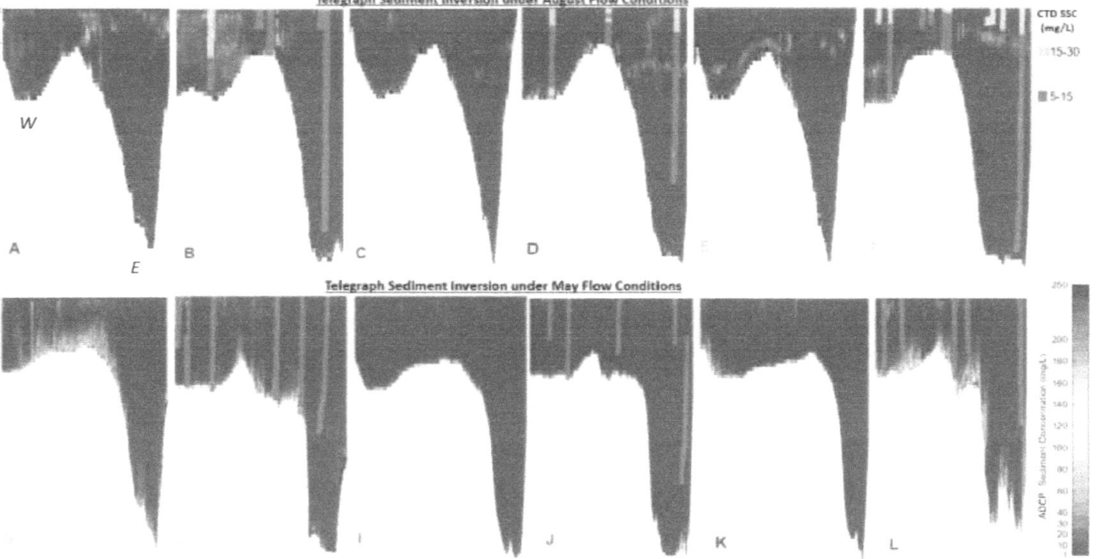

39 Figure 8Ch3: T CTDTu SSC overtop of ADCP sediment inversion SSC derived from ADCP backscatter. August ebb (purple) to flood (yellow) tide transects are presented above May flood to ebb transects. Transects A-L are from the same surveys as Figure 7Ch3 but display different data. The exact position of A-L within the tidal cycle are shown in the appendix within Figure 1Ch3Appx.

Detailed Tidal Transition Dynamics Landward within the Estuary (SDE)

Figure 9Ch3 depicts ADCP velocity direction transects during the slack tide transition at SDE in August (Figure 9Ch3A-F) and May (Figure 9Ch3G-K) and associated CTDTu salinity profiles. Under August flow conditions, the water column is highly stratified vertically throughout the slack water transition with brief horizontal stratification occurring at slack tide. Flow is mostly unidirectional during decelerating ebb flow, with small portions of flood directed flow occurring in the deep-water eastern segment and close to the bottom (Figure 9Ch3A). Velocities are highest at the surface and in the deeper parts of the channel (Table 3Ch3). Insufficient CTDTu casts were taken across the channel to assess horizontal stratification during this time. Vertically, the water column is highly stratified (VSp=1.1) with up to 5.2 PSU salinity difference between bottom and surface waters (Table 3Ch3). During slack tide, flow is flooding at depth and still ebbing at the surface (Figure 9Ch3D). Flooding velocities are highest in the deep, southern segment (Table 3Ch3). VSp indicates high vertically stratified conditions (VSp = 2.2) with more saline water at depth, while HSp indicates weak stratification across the channel (HSp = 0.6). Under accelerating flood flow, the flow is generally unidirectional (Figure 9Ch3F). Highest velocities occur mid-water column, and flow is slower above (Table 3Ch3). The horizontal stratification has disappeared (HSp = 0.2), while high vertical stratification persists (VSp=1.1).

Under low flow conditions in May, the water column experiences moderate horizontal stratification and weak vertical stratification during the high slack tide transition (Figure 9Ch3 and Table 3Ch3). During the decelerating flood (Figure 9Ch3G-H), flow is unidirectional and well mixed vertically (VSp = 0.1). Velocities are similar throughout the channel and at depth (Table 3Ch3). A moderate stratification occurs across the channel (HSp = 0.8) with the south segment up to 12.6 PSU more saline than the north. During high slack tide (Figure 9Ch3J), the north and center of the channel are already ebbing while the southern channel still experiences flood flow at depth. Weak horizontal stratification remains (HSp = 0.3), and a weak vertical stratification appears (VSp = 0.2). Under the accelerating ebb conditions (Figure 9Ch3K), the flow is unidirectional and weakly stratified in the vertical (VSp =0.2) and horizontal (HSp = 0.4). Highest velocities are concentrated at the surface in the deeper, southern segments and are higher than the flood velocities, despite similar average velocities.

Tidal current amplitude is the difference between the highest cross-channel velocity (over all seasons surveyed) and the lowest cross-channel velocity. Under the conditions surveyed in May and August (Figure 1Ch3Appx and 2Ch3Appx within the appendix) tidal current amplitude was 0.75 ms^{-1} for T (Table 2Ch3) and 0.79 ms^{-1} for SDE (Table 3Ch3). Tidal amplitude in both May and August was ~ 4 m according to Port Edward. Under HydroTrend estimated river inflows (shown in Table 2Ch3 and 3Ch3) divided by the cross-sectional area (values needed for the cross-sectional area are shown within Table 2Ch3Appx in the appendix), fluvial velocity is greater in August due to higher river inflows. Fluvial velocity and tidal current velocity are needed, along with other values displayed in Tables 2Ch3Appx, 2Ch3, and 3Ch3, to compute the dimensionless tidal (*Td)* and river (*R*) unit (shown in Tables 2Ch3 and 3Ch3). *Td* and *R* will be addressed in the discussion using Geyer (2010) estuarine classification.

13 Table 3Ch3: SDE ebb (A&H), slack (D&J), and flood (F&K) transect salinity, velocity, VSp, and HSp.
Dimensionless tidal and river unit will be addressed in the discussion along with figures from Geyer (2010). Refer to Figure 2Ch3Appx for the exact point of transects within the tidal cycle. Velocities are taken from East-West velocity vectors for SDE based on channel orientation relative to the incoming and outgoing flow. As described in the methods, Sp values 0.1 and under are well mixed, over 0.1 but under 1 are weakly stratified, and over 1 are highly stratified. The tidal mean surface is fresher and faster flowing at the surface, especially on the southern side of the channel.

SDE Variables	Units	August Ebb (A)	August Slack (D)	August Flood (F)	Tidal Mean	May Flood (H)	May Slack (J)	May Ebb (K)	Tidal Mean
HydroTrend Q (m3/s)	Q (m3/s)				1834.00				1180.00
Cross-Channel Mean Salinity	So (PSU)	4.63	1.22	5.71	3.85	16.48	19.67	15.09	15.09
Bottom Mean Salinity	Sb(PSU)	9.46	3.52	8.24	7.07	16.67	22.05	16.48	16.48
Surface Mean Salinity	Ss(PSU)	4.24	0.82	1.81	2.29	15.47	18.23	13.45	13.45
North Surface Mean Salinity	Sw(PSU)	4.29	0.78	2.45	2.51	2.8839	14.0472	17.05	17.05
South Surface Mean Salinity	Se(PSU)		1.28	2.88	2.08	15.47	18.66	11.23	11.23
Horizontal Stratification Parameter	HSp (-)		0.6	0.2	0.4	0.8	0.3	0.4	0.4
Vertical Stratification Parameter	VSp (-)	1.1	2.2	1.1	1.5	0.1	0.2	0.2	0.2
Cross-Channel Mean (E-W) Velocity	Ux (m/s)	-0.58	-0.01	0.81	NA	0.71	-0.22	-0.92	NA
Bottom Mean Velocity	Ub (m/s)	-0.29	0.15	0.43	NA	0.64	0.01	-0.71	NA
Surface Mean Velocity	Us (m/s)	-0.80	-0.14	0.89	NA	0.87	-0.42	-1.32	NA
North Surface Mean Velocity	Uw (m/s)	-0.80	-0.13	0.82	NA	0.45	-0.54	-0.72	NA
South Surface Mean Velocity	Ue (m/s)	-0.80	-0.15	0.96	NA	1.23	-0.25	-1.82	NA
Tidal Current Amplitude	U_T (m/s)				0.79				0.79
Fluvial Velocity (Q/Cross-sectional Area)	U_R (m/s)				0.044				0.028
Dimensionless Tidal Unit $U_T/(\beta g s_o A)^{1/2}$	Td (-)				1.345				0.639
Dimensionless River Unit $U_R/(\beta g s_o A)^{1/2}$	R (-)				0.075				0.023

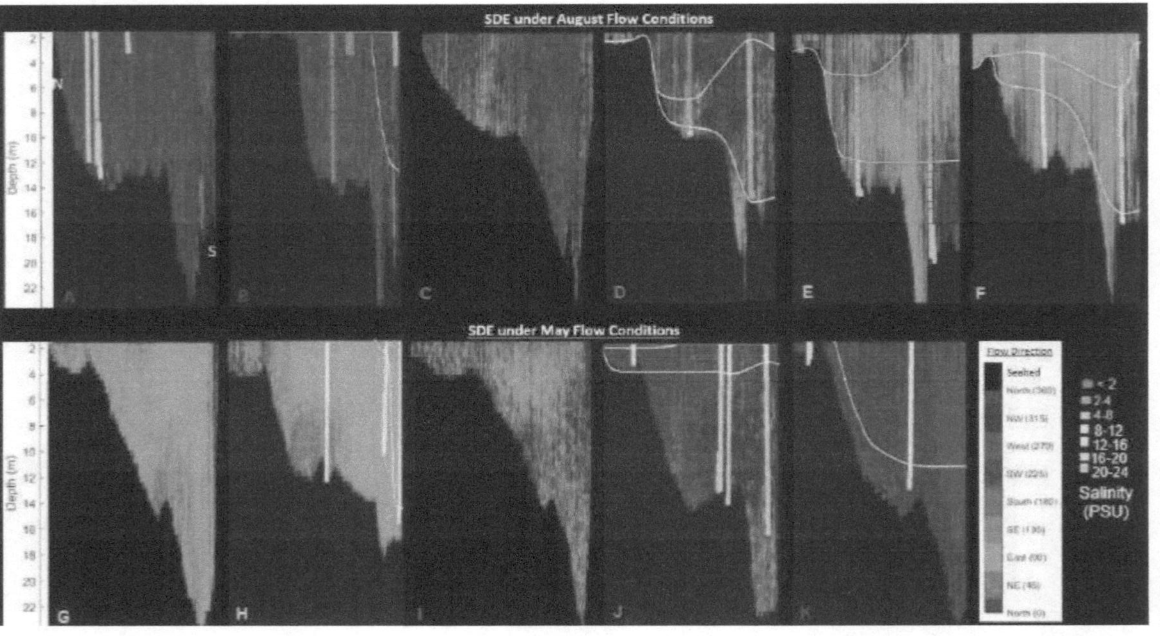

40 Figure 9Ch3: SDE flow direction and salinity casts to indicate stratification.
Salinity contours are drawn in white unless there were not enough casts taken across the channel. The exact position within the tidal cycle of transects A-K are shown in Figure 2Ch3Appx within the appendix. Ebb flow transect letters (A, B, & K) are shown in purple, slack in green, and flood in yellow. Transects were surveyed from the northern to southern bank with westward flow representing landward (flood) tidal inflow and eastern flows representing seaward (ebb) tidal outflow. August ebb to flood tide transects are presented above May flood to ebb transects.

Figure 10Ch3 displays SSC profiles (ADCP), and SSC casts (CTDTu) for the SDE May and August transects shown in Figure 9Ch3A-K. Although general trends are similar, the ADCP SSC tends to be higher at the bed and slightly lower at the surface to mid water column than the CTDTu-derived SSC casts. For example, Figure 10Ch3D, F, and J show CTDTu-derived SSC of 15-30 mgL^{-1} at the surface, while the ADCP SSC depicts 1-15 mgL^{-1}. This may be due to variations in grain size that are not captured by the ADCP backscatter inversion method; larger-than-average grains returning a stronger signal at the bed and smaller-than-average grains at the surface. The CTDTu-derived SSC has been calibrated through measured water sample SSC and should be relied upon over the ADCP inversion.

Generally, suspended material appears to be mobilized at the bed during accelerating flows and settles from the surface during decelerating flows (Figure 10Ch3 and Figure 2Ch3Appx). CTDTu casts show SSC is higher under August flow conditions with a higher river inflow. More specifically, in August during higher river inflows and high vertical stratification, highest SSC occurs across the water column in the northern, shallower segment of the channel adjacent to the tidal flats (Figure 10Ch3A). SSC remains high close to the riverbed during slack tide but decreases in the upper water column, indicating settling (Figure 10Ch3D). Areas of shear between the incoming flood and outgoing ebb tend to be more turbid. SSC increases at the bed during accelerating flood conditions while the upper water column has low SSC, indicating resuspension (Figure 10Ch3F). Highest SSC in the northern, shallow segment indicates the strongest resuspension in this region. In May during low river inflows and moderate horizontal stratification, SSC is overall less (18.51 vs 42.60 mgL^{-1} based CTDTu tidal mean values) in the water column and highest during resuspension on the accelerating ebb tide (Figure 10Ch3K). During the decelerating flood, SSC is highest in the center of the channel in the vicinity to the deep, southern segment and the northernmost bank (Figure 10Ch3G-H). During high slack tide, the water column is clearer at the bottom and less clear at the surface where ebbing is initiated (Figure 10Ch3J). Under accelerating ebb flows, high SSC near the bed predominately at the northern, shallower segment indicates resuspension (Figure 10Ch3K).

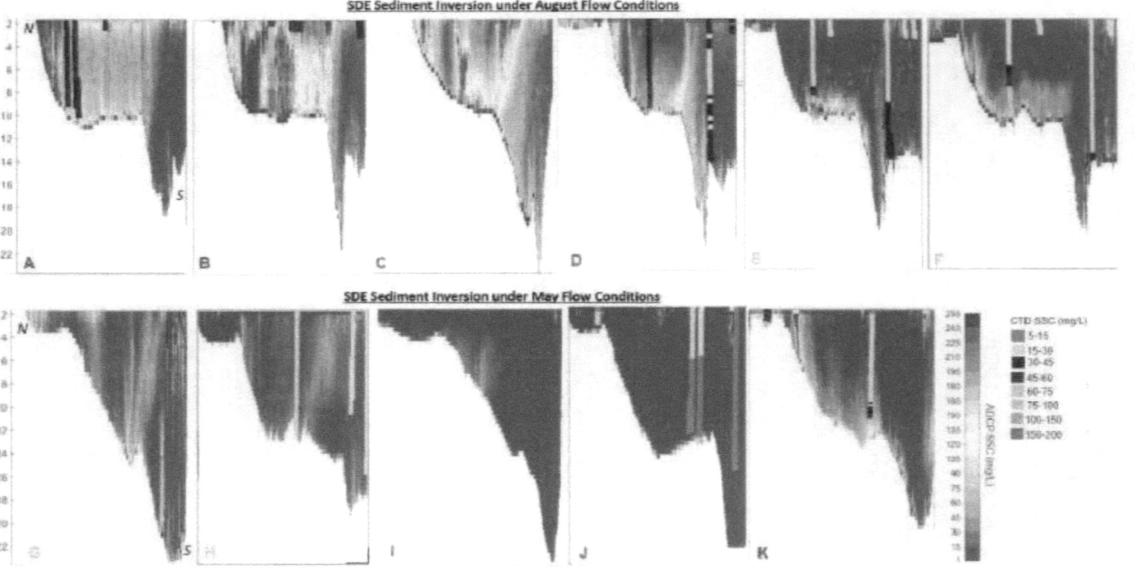

41 Figure 10Ch3: SDE CTDTu SSC overtop of ADCP sediment inversion SSC derived from ADCP backscatter. Transects A-L occur along the identical tidal cycle shown in Figure 2Ch3Appx within the appendix. Transects were surveyed from the northern to southern bank. Ebb transects letters (A, B, &K) are shown in purple, slack tide in green, and flood tide in yellow.

Tidal Prism, tidal exchange ratios, and tidal flushing

Due to the underlying assumption of well mixed conditions for tidal prism and flushing calculations (Lakhan, 2003), only sites SUE and SDE (under low flow conditions when the salinity casts are well mixed) are suitable to be used in tidal flushing and tidal prism calculations. The tidal prism between tidally drowned Skeena River mouth downstream of Ecstall and the Skeena freshwater river mouth is around 300 M m³. Due to a lack of data on river inflow without tidal interference, the following calculations will use modelled Skeena river inflow based on HydroTrend (discussed further in chapter 1) (Wild et al., in prep.). The mean annual river inflow over the tidal cycle based on the model is between 33.2 - 46.6 M m³(Wild et al., in prep.; Ch1) Thus, the ratio of riverine inflow to tidal inflow for the upper estuary is roughly 0.11-0.16.

At SDE the discharge in May at slack tide is ~10711 m^3s^{-1} with mean intertidal flood and ebb discharges between 36037.22 - 30666.61 m^3s^{-1}. This produces a tidal exchange ratio for the SDE location of 0.76. At SUE, the discharge across the channel is 9347 m^3s^{-1} at ebb tide (flood tide data is missing) and 4084 m^3s^{-1} at slack tide with an exchange ratio of 0.7. Assuming tidal conditions are similar (which they are on the day of data collection), one would expect a greater tidal exchange ratio at the seaward location with this value diminishing up the tidal prism. The slack tide discharge is ~44% of the ebb and flood tide discharge at SDE. The slack tide discharge is ~32% of the ebb tide discharge at SUE. Thus, there is a roughly ~12% decrease between slack tide and intertidal discharge every ~8km landward under ~4 m tidal amplitudes (tidal amplitudes were derived from the Port Edward tidal curve (ECCC, 2020)).

The mean percent freshwater at slack tide at SDE is 44% based on the CTDTu cast salinity and passage volume calculated using the ADCP. On the day of the May SDE transect, under HydroTrend estimated Skeena River outputs of 1197.3 m^3s^{-1} (Wild et al., in prep.). This produces an escaping volume of 59 100 000 m^3 per tidal cycle. Our August field data show stratification at SDE and can therefore not be used. On the day of data collection at SUE, HydroTrend predicts Skeena discharges of 2726 m^3s^{-1} and CTDTu casts produce a mean percent freshwater of 99.9%. This produces an escaping volume of 58 800 000 m^3 per tidal cycle. SUE was sampled under a higher river inflow than SDE. Under the same river inflows, one would expect the difference in escaping volume to be greater.

Discussion

Tidal Mean Stratification Classification of the Skeena Estuary

With a tidal prism of ~300 Mm^3 and a mean annual river inflow of 33.2 to 46.6 Mm^3, the tidal prism is under an order of magnitude greater than the river inflow indicating vertically homogenous to weakly stratified conditions in the tidal prism (Figure 1Ch3) area of the estuary. Under maximum freshet conditions, the river inflow is on the same order of magnitude as the tidal prism and could produce partially mixed conditions in the upper estuary. Inclusion of the stratified estuarine passages beyond the tidal prism zone (Figure 1Ch3) would further increase the tidal prism in proportion to the river inflow and indicate well-mixed conditions. Although well-mixed to vertically homogenous conditions are observed within the tidal prism area at SUE (Figure 4Ch3), stratification is observed at SDE onward within the estuary indicating that the tidal prism method oversimplifies the conditions within the Skeena estuary. There is also some uncertainty as to whether the tidal prism method would still be effective for the Skeena since the surface area used to calculate the prism is constrained by the bedrock. Thus, alternate classification methods have be applied over the more seaward portions of the estuary.

Further seaward from the tidal prism area, Td and R values calculated for the Skeena using tidal means over a given season (Table 2Ch3 and Table 3Ch3) have been plotted on the prognostic estuarine classification by Geyer (2010) (Figure 11Ch3). Both the seaward (T) and landward (SDE) locations under August and May flow conditions predict partially mixed conditions. Ideally, tidally, and seasonal averaged values would be used to plot and determine the general classification of the Skeena on the Geyer (2010) model. A seasonal averaged is not possible for the Skeena given the current data, and thus, the Skeena is plotted under low and moderate flows. However, based on the ECCC hydrograph (S2), August flows are closer to mean river inflows for the Skeena and may therefore be more representative of an overall mean classification for the area. Even under only two different flow conditions with low and moderate river inflows, the spread between Skeena locations spans across the partially mixed range. This highlights the challenge of defining such a dynamic area over all seasons, tides, and locations. Furthermore, flows over 1000 m^3s^{-1} more than those surveyed in the presented in the results are achieved under mean freshet conditions according to the ECCC (2020) hydrograph at Usk (Figure SK2). Thus, the classification of the Skeena under high flows is still unknown and may breach into the highly stratified estuary zone on Figure 11Ch3.

Tidal mean CTDTu cast are well-mixed to weakly stratified, which indicates that the partially stratified classification is an adequate overall description for the Skeena Estuary under moderate and low river inflows. Although select portions of the tidal cycle are highly stratified under August flows, a visual analysis of the tidal mean of CTDTu casts are weakly stratified[31]. Thus, over the entire tidal cycle mean, passages are weakly stratified. Mean salinity profiles under weak stratification are described to have a weak halocline or gradual, continuous stratification except at the bottom mixing layer that is homogenous (Valle-Levenson, 2009). Thus, based on the tidal mean of salinity casts, the estuary can be described as vertically well-mixed under low river inflow conditions with a rising tide. Under moderate river flow conditions with a falling tide, the tidal profiles are weakly horizontally stratified. Furthermore, even under local portions of the tidal cycle that produce highly stratified conditions, the surface, and the bottom of the CTDTu salinity profiles are not entirely fresh (with freshwater being 0-1 PSU) or saltwater (32-35 PSU) indication partial mixing of the water column (Figure 4Ch3). Based on the current data that show a range of high stratification (that does not produce fully fresh or saline conditions) to well mixed conditions as well as the unidirectional flows during ebb and flood, the Skeena Estuary undergoes some mixing and is therefore partially mixed.

---

[31] For example, CTD profiles are highly stratified in August in T during flood tide (Figure 4Ch3), but this changes to a weakly stratified profile during the ebb tide producing a weakly stratified tidally averaged profile.

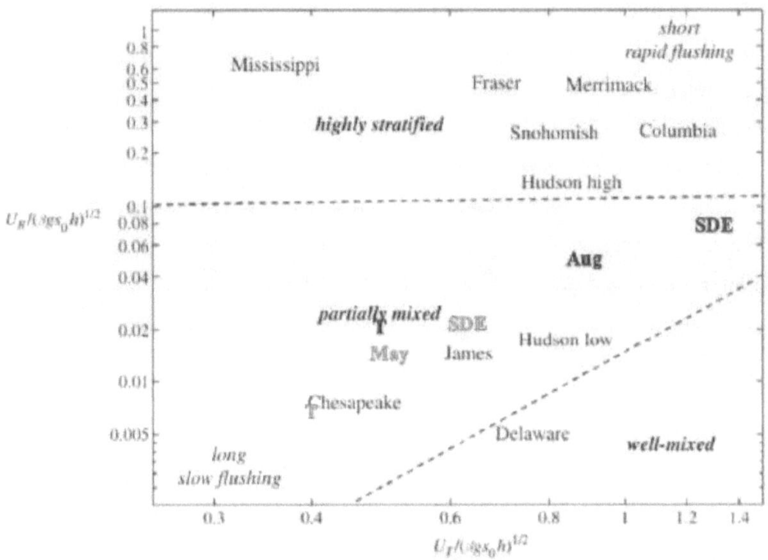

42 Figure 11Ch3: Skeena locations under May (light blue) and August (dark blue) flow conditions presented on the Geyer (2010) prognostic estuarine classification for estuaries. May and Aug labels are positioned at the mean location between T and SDE for each season. $U_T/(\beta g s_o h)^{1/2}$ on the x-axis is the formula for Td while $U_R/(\beta g s_o h)^{1/2}$ on the y-axis is the longhand for R, with values presented in Table 2Ch3 and 3Ch3 in the results.

The Geyer (2010) prognostic estuarine classification is useful to place the Skeena Estuary in relation to other estuaries based on stratification. For example, Figure 11Ch3 shows how the Skeena is not as stratified as the Fraser under the currently surveyed flow conditions. This makes sense for the Skeena estuary, as Skeena freshwater inflows are lower than that of the Fraser and the tidal range is higher. Thus, the proportion of freshwater relative to tidal strength is higher for the Fraser, allowing for the formation of a salt wedge and high stratification under particular condition (Kostaschuk and Atwood, 1990). The mean discharge at Mission is reported at 3410 m^3s^{-1} (McLean *et al.*, 1999) and Mission is approximately 85 *km* upstream of Sand Heads within the Fraser delta (Mikhailova, 2008). For comparison, the mean for the Skeena River is 2,157 m^3s^{-1} (BC Ministry of Environment, 2020). The mean tidal level fluctuates between 2-5 m depending on the season for the Fraser with a mean of 3 m (Mikhailova, 2008). In contrast, the

146

tidal level fluctuates between ~2-8 m within the Skeena with a mean level of 3.86 m (Fisheries and Oceans Canada, 2015). Furthermore, to remain under the highly stratified Geyer (2010) category under comparable tidal ranges (mean of 3.5 m) of the Skeena, the Columbia River has a mean discharge of 5210 $m^3 s^{-1}$ (Mikhailova, 2008). River discharge (mean annual freshet conditions are around 2000 $m^3 s^{-1}$ while mean conditions are around 1000 $m^3 s^{-1}$) and tidal range (~1.4 m) are lower than the Skeena for the Hudson River Estuary, but in combination produce partially mixed conditions (Geyer and Chant, 2006). Thus, although river inflows are approaching the values seen in highly stratified areas, they are still lower than the Fraser and Columbia. The higher tidal range relative to the Fraser and Columbia also increases the vertical mixing enough to reduce overall stratification and circulation, thereby creating a partially mixed estuary. Furthermore, high spatial and temporal variability, mean moderate salinity gradients with some high vertical stratification, and two-layer circulation described for the Hudson partially mixed estuary (Geyer and Chant, 2006) are also consistent with conditions observed within the Skeena Estuary.

Stratification and Flow across the Skeena Estuary

Under moderate river flow with a rising tide, stratification within the Skeena estuary occurs from below the river confluence with Ecstall (SDE) to the passages of Inverness (IV) and Telegraph (T) over the flood to ebb cycle. Meanwhile, upstream of the confluence, at SUE, the water column is well-mixed with low salinity throughout and freshwater dominance during ebb and slack. Thus, although salinity is lower prior to the estuarine divergence, SUE compared to T transects show evidence that stratification is higher seaward at T. This is opposite to the original hypothesis of increased stratification landward and prior to estuarine divergence due to a higher proportion of freshwater to seawater. Although at SDE stratification is high, at SUE, the higher proportion of freshwater and shallower depth increases mixing. Thus, as the depth and volume of water decrease and the proportion of freshwater increases landward, there must be a threshold point upon which the water column shifts from stratified to well-mixed/vertically homogenous conditions. Notice that IV, a shallow and narrow passage, expresses stratification. Thus, channel geometry and a shallow bed alone is not inducing well-mixed conditions. Rather the landward proximity of SUE prior to any channel divergence and the impact flood tide pushing freshwater landward could play a larger role in increasing the freshwater proportions and inducing well-mixed to vertically homogenous conditions under August moderate river inflows.

As observed when passage geometry and bathymetry are held constant by surveying the same location, surveys under different seasonal flows indicate that increased passage stratification induces a vertical over horizontal differentiation in the slack tide flow transition. Meanwhile, when stratification is low, other forces based on passage geometry and bathymetry, induce a horizontal transition. Cameron and Pritchard (1963) and Pritchard (1955) describe well mixed estuaries as sectionally homogenous or inhomogeneous systems where the later produces a horizontal separation of flow due to Coriolis and centrifugal forces. More specifically, on a smaller scale (i.e.: Skeena Passages), lateral (horizontal) circulation can be created by channel curvature and varying cross-lateral seabed bathymetry (Georgas and Blumberg, 2004). On a larger basin scale (i.e. Chatham Sound), transient wind effects and Coriolis force can induce lateral stratification that varies horizontally across the basin (Georgas and Blumberg, 2004). The shallow bathymetry on one side of the channel could increase friction, cause the velocities to slow, and induce the flow to breakdown and transition first over the shallow bank. Such a phenomenon is observed in SDE (Figure 9Ch3J) and T (7Ch3J) under May flows. Curves in the channel could also increase centripetal forces and allow for higher velocities in portions of the channel. When vertical stratification is high in August, flows are faster at the surface during ebb flows when the slack tide transition was observed, which could increase the velocity of the shallow bank side despite the higher friction. Thus, with a surface boost in velocity, the bottom of the channel slows and begins to turn while the entire channel surface still has the strength to outflow (Figure 7CH3C&D; Figure 9Ch3D). At some point, the stratification must overcome the passage geometry forces to induce a vertical rather than a horizontal transition. An inhomogeneous well mixed estuary is consistent with May slack tide horizontal flow conditions in the Skeena at T and vertically homogenous higher flows at SUE (Figure 4Ch3 and 6Ch3). In contrast, under August flows at T, stratification is high, and net flow is vertically bidirectional during the long slack tide transition (Figure 5Ch3). Horizontal and vertical flow transitions can cause cross-channel or cross-depth density gradients and the differential transport of saltwater across the channel.

The Skeena River inflow has a noticeable impact on salinity, stratification, and subsequent flow within the shallow (under 40 m) estuary, despite being small in comparison to the tidal prism. Under similar tidal conditions, an increase in $\sim 640\ m^3s^{-1}$ Skeena River discharge could increase percent freshwater by up to 57% at the SDE location ~28 km seaward of the freshwater Skeena River mouth (Figure 4Ch3). Thus, the current conditions and range of stratification may shift under higher river inflows.

However, under current surveyed conditions, both landward (SDE) and seaward (T), the flood tide overpowers the outgoing river inflow during higher flood flows. Contrary to the hypothesis, the flood pushes the riverine freshwater seaward producing a more salinity estuary during the flood in both T and SDE. Thus, a salt wedge was not observed further landward or within any of the transects surveyed within the estuary. Although surface velocities in T are lower during the flood due to opposing flow, the surface still changes direction and becomes unidirectional with the higher velocity flooding tide. Thus, although August T is highly stratified, it is not yet a salt wedge that would continue outflowing over top of incoming peak flood velocities at depth. Based on the Geyer (2010) classification scheme (Figure 11Ch3), a reduced tidal level fluctuation or increased river inflow could push areas of the Skeena Estuary into highly stratified salt wedge conditions. The high stratification during the flood tide and long slack tide transition during the ebb at T may indicate the potential for the emergence of salt wedge conditions under higher river inflows. However, further research is necessary to assess Skeena Estuarine conditions under freshet conditions.

Landward vs Seaward Tidal Acceleration

The exact position within the tidal cycle, based on velocity and UTC time, of a given transect within SDE is shown in the appendix in Figure 2Ch3Appx. With a comparable channel width (hydraulic geometry Table 2Ch3Appx), transect survey lengths, upon which velocity is averaged across the channel, are similar between SDE and T transects (15-25 min to survey without CTDTu casts). Thus, survey transect tidal curves should be comparable between SDE and T aside from starting continuous surveying slightly earlier or later in the day. Figure 2Ch3Appx compared to 1Ch3Appx show that acceleration and deceleration of the tide are more rapid at landward SDE than seaward T. Furthermore, ~1hr before the lowest recorded slack tide, bidirectional flow is no longer observed at SDE in August or May (Figure 2Ch3Appx; Figure 9Ch3). Meanwhile, in T, bidirectional flow can be observed 1.5-3 hr. prior to the lowest point in slack tide (Figure 1Ch3Appx; Figure 7Ch3). The difference in flow acceleration between SDE and T is the most noticeable on the decelerating flow side of the curves shown in Figures 1Ch3Appx and 2Ch3Appx. Also, August SDE values are closer compared to T to the highly stratified line and short rapid flushing portion of the Geyer (2010) classification (Figure 11Ch3). This may be indicated by a more rapid slack tide transition observed in SDE (Figure 2Ch3Appx; 9Ch3). The deeper channel at T compared to SDE may also be playing a role in the longer slack tide transition since there is a greater water volume to transition.

Suspended Sediment Flow Observations

Sediment concentrations are higher under August falling tide flows, further landward, during the ebb tide, and on the north and western sides of the channel. Concentrations are 5-15 mgL^{-1} higher at the surface landward compared to seaward even under lower river inflows in August at the landward location. At SDE, SSC is greatest throughout the water column during decelerating ebb (Figure 10Ch3B) when the flow is still dominantly unidirectional, but velocities are starting to decrease. This may be due to larger riverine sediment carried out by the ebb tide falling out of transport during the decelerating ebb tide (and being missed by the ADCP at the surface). Alternatively, the more turbid decelerating water column may be due to fine material brought in from the tidal flats, as flow recedes back into the main channel. Since, sediment is substantially higher during the ebb, this may indicate the river is a dominate sediment source at least for the SDE location and during the ebb. Tidal flats could be a source. The turbidity in May is not as strong as August (Figure 10Ch3; Figure 8Ch3). This may also indicate the influence of higher river inflow on estuarine suspended sediment. Surveys in high water in May opposed to low tide in August may be diluting some of the May sediment due to the impact of passage volume on sediment concentrations.

Contrary to the hypothesis indicating that ebb and flood would dictate if sediment were transported at the bed or in the upper water column, decelerating and accelerating tides appear to play a larger role in where sediment is transported rather than ebb and flood. At SDE and T, sediment is mobilized at the bottom during the accelerating tide (Figure 10Ch3F&K; Figure 8Ch3L) and transported throughout the mid-water-column and bottom during decelerating flows (Figure 10Ch3A&G; Figure 8Ch3A&G). Similarly, at SDE, suspended sediment during the flood tide is carried throughout the water column in May (Figure 10Ch3G&H) whereas it is mobilized at the bed in August (Figure 10Ch3E&F). Thus, the difference in the location in sediment mobilization may be due to a decelerating flood tide opposed to an accelerating flood tide in August. Furthermore, bed mobilization appears greater during accelerating ebb and flood tides while the middle water column is the most turbid during decelerating tides as velocities drop and sediment settles through the water column.

Sediment Load Divergence

Sediment load within an estuary likely consists of sediment discharge from the river, local resuspension within the estuary, and landward transport from marine sources. In the Skeena Estuary, combined sediment load (Qs) of the three passages (T, IV, M) is remarkably close to the load downstream of the last river confluence at SDE (514 vs 500 Kgs^{-1}, respectively) under low river inflow conditions. This indicates (along with the grain size fining at the bed presented in Chapter 2: Figure 3Ch2) that although tidal variation and local resuspension likely occur, overall, most of the sediment entering the estuary continues to travel through the three passages and towards the open marine areas. Our grain size samples further indicate a median grain size of suspended material of 6.9 um at SDE, a grain size that is not found in large quantities on the bed within the T or SDE area of the estuary surveyed with the ADCP (Chapter 2 Figure 3Ch2). Thus, the sediment must be sourced from elsewhere rather than solely through resuspension. Furthermore, higher suspended sediment concentrations in August compared to May low river inflows indicate a substantial river influence on suspended sediment at SDE. SDE transects display suspended sediment settling from the surface during the ebb (Figure 10Ch3 B &C) that is likely, based on the flow direction and location in the water column, riverine sourced. However, there is also some indication of higher concentrations approaching the shallow banks whereby sediment may be transferring from the tidal flats into the water column and vice versa over certain portions of the tidal cycle (Figure 10Ch3A&B). Given the recent sedimentation and descriptions on the tidal flats described in chapter two, it is also likely that suspended sediment from the river is deposited on the flats and remobilized locally throughout the tidal cycle. May suspended sediment concentrations during the flood may be from marine sediment transported upstream or settling to riverine sediment during the mixing of freshwater with seawater while flooding upstream (Figure 10CH3G&H). Satellite imagery over the Skeena estuary shows a visible plume leaving the Skeena River (Figure 1Ch2A). Thus, with all this information in mind, the river is a substantial source of suspended sediment in transport. With that said, there are still likely marine inputs that can be carried landward during storms and high tidal resuspension interacting with the seabed and water column. Therefore, although the summation of suspended load divergence between IV, T, and M are very close to SDE Qs, some differences may become more evident over different seasons and flow conditions.

Under moderate river inflows, sediment load more than doubles at the river confluence (1150 kgs^{-1}) and Telegraph Passage (430 kgs^{-1}), while Inverness passage maintains a similar load. Perhaps there are limitations on the amount of suspended sediment that can be carried into the abruptly truncating Inverness bedrock channel that prevents a marked increase in sediment load during higher river inflow conditions. However, river inflows were also the lowest during IV sampling and approaching May values than all other August passages sampled. An estimate of load through the third passage (M) under August flows is unfortunately missing.

Previous estimates of sediment load and divergence into the three passages are 25%, 37.5%, and 37.5% for Inverness, Marcus, and Telegraph, respectively (Hoos, 1975). Our low river inflow measurements seem to indicate that Marcus carries more load than Telegraph (~50% vs. ~40%). Our results also show less transport through Inverness (~10% of the incoming load) than previously estimated. Under August river inflows, percentages in IV (%5) and T (%37) differ by less than 5% lower and indicate that the percentage of sediment divergence into each channel may also vary depending on the season and flow conditions. Further studies under high river inflows would be necessary to confirm these sediment divergence estimations. Still, ADCP and CTDTu evidence from both season's surveys indicates that the divergence through Inverness has been overestimated in the past.

Conclusions

Differences in stratification and flow velocity between the May and August transects at the same location within the estuary produce horizontal (well-mixed) vs vertical (stratified) flow transitions over slack tide (e.g.: T). In addition, SUE and T slack tide flow transitions indicate that under relatively similar tidal and river inflow conditions, further landward in the estuary may be vertically homogenous (SUE) while seaward it is highly stratified (T) under select portions of the tidal cycle. Although channel geometry may increase variation in flow and stratification between the passages, the distance landward and seaward within the nearshore transects and the riverine flow relative to tidal volume range appears to play a larger role. This is evident in the differences between SDE and T with a similar channel width as well as the changes in passage flows under different seasons with the same locational geometry. The acceleration or deceleration of flow along the tidal cycle appears to impact the location of suspended sediment within the water column that is likely dominantly riverine sourced at SDE. Based on August flow conditions, that match closely to mean river inflows on the Skeena hydrograph (Figure SK3), mean tidal CTDTu casts and Geyer (2010) classifications indicate a partially stratified estuary from SDE, IV and T. Meanwhile, tidal prism calculations and ebb to slack CTDTu casts indicate well-mixed conditions from SUE landward. May to August comparisons of flow and stratification indicate dynamically varied conditions across the Skeena estuary depending on the season, tide, and river inflows.

Conclusions

High fluvial inputs (~50325.3 $M\,m^3yr^{-1}$ *of river discharge*) and macrotidal conditions (up to 7 m tides) converge within the bedrock confined fjard passages with interspersed glacial lag deposits to form a subaqueous delta and estuarine sub-environments that have shifted in comparison to unconfined models. Fluvial inputs to the estuarine system are higher than previously considered and result, in combination with the tide, in a partially mixed estuary with high variance depending on the season and location. Higher fluvial inputs are indicated by buried terrestrial material in cores, high (above 1 and below 3 $cmyr^{-1}$) sedimentation and grain size fining leading away from the landward passages, higher SSC in landward ADCP transects and CTDTu casts and under higher river inflows, and higher HydroTrend SB1 and SB2 combination results than past literature. Passages vary from vertically to horizontally stratified induced slack tide flow transitions depending on the season and seaward vs landward locations. The flow is bidirectional approaching slack tide, and unidirectional during higher ebb and flood flows. Further analysis has shown this to be an indication of partial mixing and a tide dominant influence on flow. Decreased riverine inputs later in the 21st century predicted by HydroTrend due to climate change and rising sea level will increase the accommodation space of the estuary. Furthermore, after an initial increase, passage infilling, and the tidal flat building may slow within the estuary by the end of the century. So long as the banks remain stable, these changes may provide clearer conditions to stimulate eelgrass growth but would also reduce camouflaging of salmon within the water column.

Bedrock confinement shelters select areas from waves allowing for delta plain development induced through a combination of tidal reworking and riverine infilling of the passages. The bedrock islands increase the bank areas and irregular bathymetry within the estuary, thus increasing the friction approaching the banks and centrifugal force down the passages. Varied friction promotes variations in velocity across the channel allowing for differences in sediment transport capacity throughout the water column. Overall, bedrock islands and confinement increase variation in the dissemination of marine and fluvial processes across the estuary. These variations allow for a patchwork tripartite structure of sub-environments at the seabed with large variations in stratification and flow depending on estuarine inputs.

References

Aberle, J., Rennie, C., Admiraal, D., & Muste, M. (Eds.). (2017). *Experimental Hydraulics: Methods, Instrumentation, Data Processing and Management: Volume II: Instrumentation and Measurement Techniques.* CRC Press.

Allen, G. P., Smith, D. G., Reinson, G. E., Zaitlin, B. A., & Rahmani, R. A. (1991). Clastic tidal sedimentology.

Algeo, T. J., Phillips, M., Jaminski, J., & Fenwick, M. (1994). High-resolution X-radiography of laminated sediment cores. *Journal of Sedimentary Research, 64*(3a), 665-668.

Attard, M. E., Venditti, J. G., & Church, M. (2014). Suspended sediment transport in Fraser River at Mission, British Columbia: New observations and comparison to historical records. *Canadian Water Resources Journal/Revue canadienne des ressources hydriques, 39*(3), 356-371.

Baek, Y. S., Lee, S. H., & Chang, T. S. (2017). Last interglacial to Holocene sedimentation on intertidal to subtidal flats revealed by seismic and deep-core sediment analyses, southwest coast of Korea. *Quaternary International, 459*, 45-54.

Battin, T. J., Kaplan, L. A., Findlay, S., Hopkinson, C. S., Marti, E., Packman, A. I., ... & Sabater, F. (2008). Biophysical controls on organic carbon fluxes in fluvial networks. *Nature Geoscience, 1*(2), 95

Bauer, J. E., Cai, W. J., Raymond, P. A., Bianchi, T. S., Hopkinson, C. S., & Regnier, P. A. (2013). The changing carbon cycle of the coastal ocean. *Nature, 504*(7478), 61.

Becker, M., Maushake, C., & Winter, C. (2018). Observations of mud-induced periodic stratification in a hyperturbid estuary. *Geophysical Research Letters.*

Benke, A., Cushing C. (2009). Field Guide to River of North America. *Academic Press.* ISBN 978-0-12-378577-0.

Bianchi, T. S., & Allison, M. A. (2009). Large-river delta-front estuaries as natural "recorders" of global environmental change. *Proceedings of the National Academy of Sciences, 106*(20), 8085-8092.

Bierman, P. R., & Montgomery, D. R. (2014). Key concepts in geomorphology (p. 494).

Binda, G. G., Day, T. J., & Syvitski, J. P. M. (1986). *Terrestrial sediment transport into the marine environment of Canada. Annotated bibliography and data: Environ. Canada, Sediment. Survey Section Rept.* IWD-HQ-WRB-SS-86-1.

Boothroyd, J. C. (1978). Mesotidal inlets and estuaries. In *Coastal sedimentary environments* (pp. 287-360). Springer, New York, NY.

Boyd, R., Dalrymple, R., & Zaitlin, B. A. (1992). Classification of clastic coastal depositional environments. *Sedimentary Geology, 80*(3-4), 139-150.

Bradley, R. W., Venditti, J. G., Kostaschuk, R. A., Church, M., Hendershot, M., & Allison, M. A. (2013). Flow and sediment suspension events over low-angle dunes: Fraser Estuary, Canada. *Journal of Geophysical Research: Earth Surface, 118*(3), 1693-1709.

British Columbia Ministry of Natural Gas Development, (ND). *LNG 101: A Guide to British Columbia's Liquified Natural Gas Sector* (Report). British Columbia Ministry of Natural Gas Development. Retrieved 2018 from: http://www.gov.bc.ca/mngd/doc/LNG101.pdf

British Columbia Ministry of Environment (2020). Normal Runoff from British Columbia- Study 406. Government of British Columbia. Retrieved from http://www.env.gov.bc.ca/wsd/plan_protect_sustain/groundwater/library/bc-runoff.html

British Columbia Ministry of Environment (2016). Indicators of Climate Change for British Columbia. 2016 Update. Retrieved May 2020 from https://www2.gov.bc.ca/assets/gov/environment/research-monitoring-and-reporting/reporting/envreportbc/archived-reports/climate-change/climatechangeindicators-13sept2016_final.pdf

British Columbia ShoreZone Map (1998-2000). ShoreZone Geology. Retrieved from the Strait of Georgia Data Center in 2020: http://sogdatacentre.ca/interactive-map/shorezone_map/shorezone_map.html

Bronk Ramsey, C. (2017). Methods for Summarizing Radiocarbon Datasets. *Radiocarbon, 59*(2), 1809-1833.

Brouwer, R. L., Schramkowski, G. P., Dijkstra, Y. M., & Schuttelaars, H. M. (2018). Time evolution of estuarine turbidity maxima in well-mixed, tidally dominated estuaries: The role of availability- and erosion-limited conditions. *Journal of Physical Oceanography, 48*(8), 1629-1650.

Cameron, W. M., & Pritchard, D. W. (1963). Estuaries In: MN Hill (editor): The Sea John Wiley and Sons. *New York, 2*, 306-324.

Carr-Harris, C., Gottesfeld, A. S., & Moore, J. W. (2015). Juvenile salmon usage of the Skeena River estuary. *PlOS one, 10*(3), e0118988

Cavalcante G.H. (2016) Well-Mixed Estuary. In: Kennish M.J. (eds) Encyclopedia of Estuaries. Encyclopedia of Earth Sciences Series. Springer, Dordrecht. https://doi.org/10.1007/978-94-017-8801-4_184

Chen, Z., Wang, Z., Liu, Y., Wang, S., & Leng, C. (2016). Estimating the flow velocity and discharge of ADCP unmeasured area in tidal reach. *Flow Measurement and Instrumentation, 52*, 208-218.

Clague, J. J. (1984). *Quaternary geology and geomorphology, Smithers-Terrace-Prince Rupert area, British Columbia* (Vol. 413). Geological Survey of Canada.

Conway, K. W.; Bornhold, B. D.; Barrie, J.V. (1996). Surficial geology and sedimentary processes, Skeena River delta, *Geological Survey of Canada*, Current Research no. 1996-E, 1996 p. 23-32, doi: 10.4095/207870

Crawford, M. L., Hollister, L. S., & Woodsworth, G. J. (1987). Crustal deformation and regional metamorphism across a terrane boundary, Coast Plutonic Complex, British Columbia. *Tectonics, 6*(3), 343-361.

Chézy, A. (1776). Formula to find the uniform velocity that the water will have in a ditch or in a canal of which the slope is known. *Journal Association of Engineering Societies, 18*.

CSDMS contributors (2018). *Model: HydroTrend.* Retrieved 2018 from: https://csdms.colorado.edu/mediawiki/index.php?title=Model:HydroTrend&oldid=216854

Cui, B. L., & Li, X. Y. (2011). Coastline change of the Yellow River estuary and its response to the sediment and runoff (1976–2005). *Geomorphology, 127*(1-2), 32-40.

Dalrymple, R. W., Zaitlin, B. A., & Boyd, R. (1992). Estuarine facies models; conceptual basis and stratigraphic implications. *Journal of Sedimentary Research, 62*(6), 1130-1146.

Dalrymple, R. W., Mackay, D. A., Ichaso, A. A., & Choi, K. S. (2012). Processes, morphodynamics, and facies of tide-dominated estuaries. In *Principles of tidal sedimentology* (pp. 79-107). Springer, Dordrecht.

Day, J. H. (1980). What is an estuary. *South African Journal of Science, 76*(5), 198-198.

De Blij, H. J., Muller, P. O., & Williams, R. S. (2004). *Physical geography: the global environment* (Vol. 2). Oxford University Press.

De Groot, A. Review of the Hydrology, Geomorphology, Ecology and Management of the Skeena River Floodplain Adrian de Groot, MSc, RPBio.

Deines, K. L. (1999, March). Backscatter estimation using broadband acoustic Doppler current profilers. In *Proceedings of the IEEE Sixth Working Conference on Current Measurement (Cat. No. 99CH36331)* (pp. 249-253). IEEE.

Dinehart, R. L., & Burau, J. R. (2005). Repeated surveys by acoustic Doppler current profiler for flow and sediment dynamics in a tidal river. *Journal of hydrology, 314*(1-4), 1-21.

Downing, A., Thorne, P. D., & Vincent, C. E. (1995). Backscattering from a suspension in the near field of a piston transducer. *The Journal of the Acoustical Society of America, 97*(3), 1614-1620.

Dugan, J. E., Airoldi, L., Chapman, M. G., Walker, S. J., Schlacher, T., Wolanski, E., & McLusky, D. (2011). 8.02-Estuarine and coastal structures: environmental effects, a focus on shore and nearshore structures. *Treatise on estuarine and coastal science, 8,* 17-41.

Duxbury, A. C., Duxbury, A. B., & Sverdrup, K. A. The Oceans of the World, PWN Warsaw 2002.

Environment and Climate Change Canada (Updated 2020). [Historical Hydrometric Data, Skeena River at Usk Station]. Raw data. Retrieved 2020 from https://wateroffice.ec.gc.ca/mainmenu/historical_data_index_e.html

(ECCC) Environment and Climate Change Canada (2020). Canadian Climate Normal 1981-2010 Station Data. Prince Rupert Climate ID: 1066481.

Enkin, R. J., Dallimore, A., Baker, J., Southon, J. R., & Ivanochko, T. (2013). A new high-resolution radiocarbon Bayesian age model of the Holocene and Late Pleistocene from core MD02-2494 and others, Effingham Inlet, British Columbia, Canada; with an application to the paleoseismic event chronology of the Cascadia Subduction Zone. *Canadian Journal of Earth Sciences, 50*(7), 746-760.

Fairbridge, R. W., & RW, F. (1980). The estuary: its definition and geodynamic cycle.

Fisheries and Oceans Canada (2019) Prince Rupert (#9354) Tidal Predictions. Retrieved from https://www.tides.gc.ca/eng/station?sid=9354

Fisheries and Oceans Canada (2015). *Technical Review of 3D Modelling of Potential Effects of Marine Structures on Site Hydrodynamics and sedimentation from the construction of the Pacific Northwest Liquefied Natural Gas Terminal* (Report). Pacific Region, Canadian Science Advisory Secretariat Science Response 2015/027, Fisheries and Oceans Canada, Nanaimo, BC. Retrieved from: http://publications.gc.ca/pub?id=9.803009&sl=0.

Flett Research LTD. (2020). Flett Research Radioisotopic Services. Accessed from: http://www.flettresearch.ca/Radioisotope.html

Flemming, B. W. (2011). 3.02 Geology, Morphology, and Sedimentology of Estuaries and Coasts. *Treatise on Estuarine and Coastal Science. Academic Press, Waltham*, 7-38.

Folk, R. L., & Ward, W. C. 1957. Brazos River Bar: A study in the significance of grain size parameters. Journal of Sedimentary Petrology, 27(1): 3-26.

Fontes R.F.C., Miranda L.B., Andutta F. (2016) Estuarine Circulation. In: Kennish M.J. (eds) Encyclopedia of Estuaries. Encyclopedia of Earth Sciences Series. Springer, Dordrecht. https://doi.org/10.1007/978-94-017-8801-4_176

Francois, R. E., & Garrison, G. R. (1982). Sound absorption based on ocean measurements: Part I: Pure water and magnesium sulfate contributions. *The Journal of the Acoustical Society of America, 72*(3), 896-907.

Freshwater Fisheries Society of BC (2006). *Skeena Feature Lake Guide.* Go Fish BC, British Columbia Ministry of Environment. Retrieved 2018 from: http://www.env.gov.bc.ca/skeena/fish/R6%20featured%20lakes%20guide.pdf

Friedman, G. M., & Sanders, J. E. (1978). *Principles of sedimentology.* Wiley.

Fulton R. (1995). Surficial Material of Canada, scale 1:5 000 000 [Map 1880A]. *Geological Survey of Canada.* Retrieved from https://www.nrcan.gc.ca/earth-sciences/geography/atlas-canada/selected-thematic-maps/16876#surficialmaterialsandglaciation

Gales, J. A., Talling, P. J., Cartigny, M. J., Hughes Clarke, J., Lintern, G., Stacey, C., & Clare, M. A. (2019). What controls submarine channel development and the morphology of deltas entering deep-water fjords?. *Earth Surface Processes and Landforms, 44*(2), 535-551.

Galloway, W. E. (1975). Process framework for describing the morphologic and stratigraphic evolution of deltaic depositional systems.

Gartner, J. W. (2004). Estimating suspended solids concentrations from backscatter intensity measured by acoustic Doppler current profiler in San Francisco Bay, California. *Marine Geology, 211*(3-4), 169-187.

Gehrels, G., Rusmore, M., Woodsworth, G., Crawford, M., Andronicos, C., Hollister, L., ... & Davidson, C. (2009). U-Th-Pb geochronology of the Coast Mountains batholith in north-coastal British Columbia: Constraints on age and tectonic evolution. *Geological Society of America Bulletin, 121*(9-10), 1341-1361.

GeoBC (2012, February 20). [Digital Elevation Model for British Columbia - CDED - 1:250,000]. Open Government License-British Columbia Data. Retrieved from: https://catalogue.data.gov.bc.ca/dataset/digital-elevation-model-for-british-columbia-cded-1-250-000

Georgas, N., & Blumberg, A. F. (2004). The influence of centrifugal and Coriolis forces on the circulation in a curving estuary. In *Estuarine and Coastal Modeling (2003)* (pp. 541-558).

Geyer, W. R., & Chant, R. (2006). *The physical oceanography processes in the Hudson River estuary* (pp. 13-23). New York: Cambridge University Press.

Geyer, W. R. (2010). Estuarine salinity structure and circulation. *Contemporary issues in estuarine physics*, 12-26.

Giorgi, F. and Francisco, R., 2000: Evaluating uncertainties in the prediction of regional climate change. Geophysical Research Letters, 27(9), 1295-1298.

Goldfish (2006). *Soil Texture and Structure.* Retrieved 2020 from
https://www.esf.edu/for/briggs/FOR345/labtxt03.htm

Gottesfeld, A., & Rabnett, K. A. (2008). *Skeena river fish and their habitat.* Portland, OR; Hazelton, BC;
Ecotrust.

Gordon, R. (1996). Acoustic Doppler current profiler: principles of operation, a practical primer. for
Broadband ADCPs. *RD Instruments, San Diego, California.*

Green, M.O., 2011. Dynamics of very small waves and associated sediment resuspension on an estuarine
intertidal flat. Estuarine, Coastal and Shelf Science, 93(4): 449–459.

Greenwood, K. C. (1991). Performance of an acoustic doppler current profiler in Knight Inlet, British
Columbia. *Msc book Royal Roads Military College.*

Grobe, H., Winn, K., Werner, F., Driemel, A., Schumacher, S., & Sieger, R. (2017). The GIK-Archive of
sediment core radiographs with documentation. *Earth System Science Data, 9*, 969-976.

Gulf of Georgia Cannery Society (2020). Inverness Canneries. Retrieved 2020 from
http://tidestotins.ca/cannery/inverness/

Hage, S., Cartigny, M. J., Clare, M. A., Sumner, E. J., Vendettuoli, D., Hughes Clarke, J. E., ... & Englert,
R. G. (2018). How to recognize crescentic bedforms formed by supercritical turbidity currents in
the geologic record: Insights from active submarine channels. *Geology, 46*(6), 563-566.

Hamblin, P. F., Lum, K. R., Comba, M. E., & Kaiser, K. L. E. (1988). Observations of suspended
sediment flux over a tidal cycle in the region of the turbidity maximum of the upper St. Lawrence
River Estuary. In *Hydrodynamics and Sediment Dynamics of Tidal Inlets* (pp. 245-256). Springer,
New York, NY.

Hamilton, T. S., Enkin, R. J., Riedel, M., Rogers, G. C., Pohlman, J. W., & Benway, H. M. (2015).
Slipstream: an early Holocene slump and turbidite record from the frontal ridge of the Cascadia
accretionary wedge off western Canada and paleoseismic implications. *Canadian Journal of
Earth Sciences, 52*(6), 405-430.

Hansen, D. V., & Rattray Jr, M. (1966). New dimensions in estuary classification 1. *Limnology and
Oceanography, 11*(3), 319-326.

Hausfather, Z., & Peters, G. P. (2020). Emissions–the 'business as usual story is misleading.

Hayes, M. O. (1975). Morphology of Sand Accumulations in Estuaries. In, CRONIN, LE (ed.), Estuarine
Research. *New York, Academic Press, 2*, 3-22.

Hibma, A., De Vriend, H. J., & Stive, M. J. F. (2003). Numerical modelling of shoal pattern formation in
well-mixed elongated estuaries. *Estuarine, Coastal and Shelf Science, 57*(5-6), 981-991.

Hizzett, J. L., Hughes Clarke, J. E., Sumner, E. J., Cartigny, M. J. B., Talling, P. J., & Clare, M. A.
(2018). Which triggers produce the most erosive, frequent, and longest runout turbidity currents
on deltas?. *Geophysical Research Letters, 45*(2), 855-863.

Hjulstrom, F. (1935). Studies of the morphological activity of rivers as illustrated by the river fyris,
bulletin. *Geological Institute Upsalsa, 25*, 221-527.

Hoos, L. M. (1975). *The Skeena River Estuary, Status of Environmental Knowledge to 1975: Report of
the Estuary Working Group, Department of the Environment* (Report). Regional Board, Pacific
Region (No. 3). Environment Canada.

Hu, K., Ding, P., Wang, Z., & Yang, S. (2009). A 2D/3D hydrodynamic and sediment transport model for the Yangtze Estuary, China. *Journal of Marine Systems, 77*(1-2), 114-136.

Hutchison, W. (1982). *Geology of the Prince Rupert-Skeena Map Area, British Columbia* (Report). Geological Survey of Canada, Memoir 394, 1982, 116 pages, DOI: https://doi.org/10.4095/116164

Hutton, E. W. H., & Syvitski, J. P. M. (2008). Sedflux 2.0: An advanced process-response model that generates three-dimensional stratigraphy. *Computers and Geosciences, 34*(10), 1319-1337. doi:10.1016/j.cageo.2008.02.013

IPCC, 2013: Summary for Policymakers. In: Climate Change 2013: The Physical Science Basis. Contribution of Working Group I to the Fifth Assessment Report of the Intergovernmental Panel on Climate Change [Stocker, T.F., D. Qin, G.-K. Plattner, M. Tignor, S.K. Allen, J. Boschung, A. Nauels, Y. Xia, V. Bex and P.M. Midgley (eds.)]. Cambridge University Press, Cambridge, United Kingdom and New York, NY, USA.

IPCC, 2014: Climate Change 2014: Synthesis Report. Contribution of Working Groups I, II and III to the Fifth Assessment Report of the Intergovernmental Panel on Climate Change [Core Writing Team, R.K. Pachauri and L.A. Meyer (eds.)]. IPCC, Geneva, Switzerland, 151 pp.

Isla, F. I., & Bujalesky, G. G. (2004). Morphodynamics of a gravel-dominated macrotidal estuary: Rio Grande, Tierra del Fuego. *Revista de la Asociación Geológica Argentina, 59*(2), 220-228.

James, T S; Henton J A; Leonard, LJ; Darlingon, A; Forbes D L; Craymer, M (2015). Tabulated values of relative sea-level projections in Canada and the adjacent mainland United States. Geological Survey of Canada, Open File 7942, 81 pages, https://doi.org/10.4095/297048

Jay, D. A., Talke, S. A., Hudson, A., & Twardowski, M. (2015). Estuarine turbidity maxima revisited: Instrumental approaches, remote sensing, modeling studies, and new directions. In *Developments in sedimentology* (Vol. 68, pp. 49-109). Elsevier.

Ketchum, B. H., & Rawn, A. M. (1951). The Flushing of Tidal Estuaries [with Discussion]. *Sewage and Industrial Wastes*, 198-209.

Kettner, A.J., and Syvitski, J.P.M., (2008). HydroTrend version 3.0: A Climate-Driven Hydrological Transport Model that Simulates Discharge and Sediment Load leaving a River System. *Computers & Geosciences, 34*(10), 1170-1183.

Kilham, N. E., Roberts, D., & Singer, M. B. (2012). Remote sensing of suspended sediment concentration during turbid flood conditions on the Feather River, California—a modeling approach. *Water Resources Research, 48*(1).

Kostaschuk, R. A., Church, M. A., & Luternauer, J. L. (1989). Bedforms, bed material, and bedload transport in a salt-wedge estuary: Fraser River, British Columbia. *Canadian Journal of Earth Sciences, 26*(7), 1440-1452.

Kostaschuk, R. A., & Luternauer, J. L. (1989). The role of the salt-wedge in sediment resuspension and deposition: Fraser River estuary, Canada. *Journal of Coastal Research*, 93-101.

Kostaschuk, R. A., & Atwood, L. A. (1990). River discharge and tidal controls on salt-wedge position and implications for channel shoaling: Fraser River, British Columbia. *Canadian Journal of Civil Engineering, 17*(3), 452-459.

Larsen, C. F., Motyka, R. J., Freymueller, J. T., Echelmeyer, K. A., & Ivins, E. R. (2004). Rapid uplift of southern Alaska caused by recent ice loss. *Geophysical Journal International, 158*(3), 1118-1133.

Lee, K. S., Park, S. R., & Kim, J. B. (2005). Production dynamics of the eelgrass, Zostera marina in two bay systems on the south coast of the Korean peninsula. *Marine Biology, 147*(5), 1091-1108.

Lee, H. J., Park, J. Y., Lee, S. H., Lee, J. M., & Kim, T. K. (2013). Suspended sediment transport in a rock-bound, macrotidal estuary: Han Estuary, Eastern Yellow Sea. *Journal of Coastal Research, 29*(2), 358-371.

Leopold, L. B., & Maddock, T. (1953). *The hydraulic geometry of stream channels and some physiographic implications* (Vol. 252). US Government Printing Office.

Levin, L. A., Boesch, D. F., Covich, A., Dahm, C., Erséus, C., Ewel, K. C., ... & Strayer, D. (2001). The function of marine critical transition zones and the importance of sediment biodiversity. *Ecosystems, 4*(5), 430-451.

Li, M., Chen, Z., Yin, D., Chen, J., Wang, Z., & Sun, Q. (2011). Morphodynamic characteristics of the dextral diversion of the Yangtze River mouth, China: tidal and the Coriolis Force controls. *Earth Surface Processes and Landforms, 36*(5), 641-650.

Li, M., Ge, J., Kappenberg, J., Much, D., Nino, O., & Chen, Z. (2014). Morphodynamic processes of the Elbe River estuary, Germany: the Coriolis effect, tidal asymmetry and human dredging. *Frontiers of earth science, 8*(2), 181-189.

Liangqing, X., & Galloway, W. E. (1991). Fan-Delta, Braid Delta and the Classification of Delta Systems. *Acta Geologica Sinica-English Edition, 4*(4), 387-400.

Lintern, D.G. and Haaf, J. (2014). Modeling the Mackenzie River Basin: Current Conditions and Climate Change Scenarios. *Geological Survey of Canada*, Open File 5531. doi:10.4095/293313. Retrieved from: http://geoscan.nrcan.gc.ca/starweb/geoscan/servlet.starweb?path=geoscan/fulle.web&search1=R= 293313

Lotze, H. K., Lenihan, H. S., Bourque, B. J., Bradbury, R. H., Cooke, R. G., Kay, M. C., ... & Jackson, J. B. (2006). Depletion, degradation, and recovery potential of estuaries and coastal seas. *Science, 312*(5781), 1806-1809.

Luternauer, J.L. (1976). Skeena Delta Sedimentation, British Coumbia. In *Geological Survey of Canada Paper no.76-1A*. doi: https://doi.org/10.4095/119844

Maan, D. C., van Prooijen, B. C., & Wang, Z. B. (2019). Progradation Speed of Tide-Dominated Tidal Flats Decreases Stronger Than Linearly With Decreasing Sediment Availability and Linearly With Sea Level Rise. *Geophysical Research Letters, 46*(1), 262-271.

Mcgrath, D., Sass, L., O'Neel, S., Arendt, A., & Kienholz, C. (2017). Hypsometric control on glacier mass balance sensitivity in Alaska and northwest Canada. *Earth's Future, 5*(3), 324-336.

McKenna, G. T., Luternauer, J. L., & Kostaschuk, R. A. (1992). Large-scale mass-wasting events on the Fraser River delta front near Sand Heads, British Columbia. *Canadian Geotechnical Journal, 29*(1), 151-156.

McLaren, P. (2016). The environmental implications of sediment transport in the waters of Prince Rupert, British Columbia, Canada: a comparison between kinematic and dynamic approaches. *Journal of Coastal Research, 32*(3), 465-482.

McLean, D. G., Church, M., & Tassone, B. (1999). Sediment transport along lower Fraser River: 1. Measurements and hydraulic computations. *Water Resources Research, 35*(8), 2533-2548.

Mikhailova, M. (2008). Hydrological and Morphological Features of River Mouths of Different Types (the Columbia Estuary and the Fraser Delta as Examples). *Environmental Research, Engineering & Management, 46*(4).

Millero, F. J. (1986). The pH of estuarine waters. *Limnology and Oceanography, 31*(4), 839-847.

Milliman, J. D., & Syvitski, J. P. (1992). Geomorphic/tectonic control of sediment discharge to the ocean: the importance of small mountainous rivers. *The Journal of Geology, 100*(5), 525-544.

Moate, B. D., & Thorne, P. D. (2012). Interpreting acoustic backscatter from suspended sediments of different and mixed mineralogical composition. *Continental Shelf Research, 46*, 67-82.

Moore, K. A., Wetzel, R. L., & Orth, R. J. (1997). Seasonal pulses of turbidity and their relations to eelgrass (Zostera marina L.) survival in an estuary. *Journal of Experimental Marine Biology and Ecology, 215*(1), 115-134.

Moradi, G., Vermeulen, B., Rennie, C. D., Cardot, R., & Lane, S. N. (2019). Evaluation of aDcp processing options for secondary flow identification at river junctions. *Earth Surface Processes and Landforms, 44*(14), 2903-2921.

Mulder, T., & Syvitski, J. P. (1995). Turbidity currents generated at river mouths during exceptional discharges to the world oceans. *The Journal of Geology, 103*(3), 285-299.

Natural Resource Canada (NRCAN) (2016, January 25). [National Hydro Network (NHN)- Geobase Series]. Open Government License-British Columbia Data. Retrieved from: https://open.canada.ca/data/en/dataset/a4b190fe-e090-4e6d-881e-b87956c07977

Natural Resources Canada Expedition Database (2019). Recent Additions to the Expedition Database. Accessed 2020 from https://ed.gdr.nrcan.gc.ca/whatsnew_e.php

Norbit (ND). Products. Retrieved 2020 from https://norbit.com/subsea/products/

Northwest Institute & SkeenaWild Conservation Trust (2018). *LNG in Northern BC, Proposed Projects.* Retrieved 2018, from http://lnginnorthernbc.ca/index.php/proposed-projects/

Ocean Ecology (2014). *Skeena Estuary Juvenile Salmon Habitat.* Retrieved 2018 from: http://skeenawild.org/images/uploads/docs/Skeena_River_Estuary_Juvenile_Salmon_Habitat.pdf

Orton, G. J., & Reading, H. G. (1993). Variability of deltaic processes in terms of sediment supply, with particular emphasis on grain size. *Sedimentology, 40*(3), 475-512.

Pasternack, G. B., & Brush, G. S. (1998). Sedimentation cycles in a river-mouth tidal freshwater marsh. *Estuaries, 21*(3), 407-415.

Paturej, E. (2008). Estuaries-types, role and impact on human life. *Baltic Coastal Zone. Journal of Ecology and Protection of the Coastline, 12.*Skinner, C. J., Coulthard, T. J., Parsons, D. R., Ramirez, J. A., Mullen, L., & Manson, S. (2015). Simulating tidal and storm surge hydraulics with a simple 2D inertia based model, in the Humber Estuary, UK. *Estuarine, Coastal and Shelf Science, 155*, 126-136.

PCIC, 2013. Statistically Downscaled Climate Scenarios. Website retrieved from: https://www.pacificclimate.org/data/statistically-downscaled-climate-scenarios

Perillo, G. M. (1995). Definitions and geomorphologic classifications of estuaries. In *Developments in Sedimentology* (Vol. 53, pp. 17-47). Elsevier.

Phillips, S. R., & Costa, M. (2017). Spatial-temporal bio-optical classification of dynamic semi-estuarine waters in western North America. *Estuarine, Coastal and Shelf Science, 199*, 35-48.

Prandle, D. (2009). *Estuaries: dynamics, mixing, sedimentation and morphology*. Cambridge University Press.

Pritchard, D. W. (1952). Estuarine hydrography. In *Advances in geophysics* (Vol. 1, pp. 243-280). Elsevier.

Pritchard D.W., 1955. Estuarine circulation patterns. Proc. Amer. Soc. Civil Engrs., 81, 717.

Pritchard D.W., 1967. What is an estuary: physical viewpoint. In: Estuaries. (Ed.) G.H. Lauff, Amer. Assoc. Adv. Sci. Publ., Washington, D.C., 83, 3-5.

Quality Positioning Services (2020). Qimera. Retrieved 2020 from https://qps.nl/qimera/

Reimer, P. J., Bard, E., Bayliss, A., Beck, J. W., Blackwell, P. G., Bronk Ramsey, C., Grootes, P. M., Guilderson, T. P., Haflidason, H., Hajdas, I., HattŽ, C., Heaton, T. J., Hoffmann, D. L., Hogg, A. G., Hughen, K. A., Kaiser, K. F., Kromer, B., Manning, S. W., Niu, M., Reimer, R. W., Richards, D. A., Scott, E. M., Southon, J. R., Staff, R. A., Turney, C. S. M., & van der Plicht, J. (2013). IntCal13 and Marine13 Radiocarbon Age Calibration Curves 0-50,000 Years cal BP. *Radiocarbon, 55*(4).

Rennie, C. D., & Rainville, F. (2006). Case study of precision of GPS differential correction strategies: Influence on aDcp velocity and discharge estimates. *Journal of hydraulic engineering, 132*(3), 225-234.

Rennie, C. D., Millar, R. G., & Church, M. A. (2002). Measurement of bed load velocity using an acoustic Doppler current profiler. *Journal of Hydraulic Engineering, 128*(5), 473-483.

Richardson, J. S., & Milner, A. M. (2005). Pacific coast rivers of Canada and Alaska. In A.C. Benke & C.E.Cushing (Eds.), *Rivers of North America* (pp. 734-773). London, UK: Elsevier.Ritter, D. F., Kochel, R. C., & Miller, J. R. (2002). Process geomorphology. 4th Ed. McGraw-Hill, Boston, MA.

Ritter, D. F. (1978). Process Geomorphology, Chapter 7 Fluvial Landforms, Deltas. *Dubuque, IA: Wm. C. Brown, 539*.

Röttgers, R., Heymann, K., & Krasemann, H. (2014). Suspended matter concentrations in coastal waters: Methodological improvements to quantify individual measurement uncertainty. *Estuarine, Coastal and Shelf Science, 151*, 148-155.

RGI (2017). [Randolph Glacier Inventory 6.0 – A Dataset of Global Glacier Outlines: Version 6.0] Raw data & Technical Report. Global Land Ice Measurements from Space (GLIMS), Colorado, USA. Digital Media. DOI: https://doi.org/10.7265/N5-RGI-60

Sassi, M. G., Hoitink, A. J. F., & Vermeulen, B. (2012). Impact of sound attenuation by suspended sediment on ADCP backscatter calibrations. *Water Resources Research, 48*(9).

Schartau, M., Riethmüller, R., Flöser, G., van Beusekom, J. E. E., Krasemann, H., Hofmeister, R., & Wirtz, K. (2019). On the separation between inorganic and organic fractions of suspended matter in a marine coastal environment. *Progress in Oceanography, 171*, 231-250.

Schiefer, E., Reid, K., Burt, A., & Luce, J. (2001). *Assessing Natural Sedimentation Patterns and Impacts of Land Use on Sediment Yield: A Lake-sediment–based Approach*. In DAA Toews and S. Chatwin (Eds.), Watershed assessment in the Southern Interior of British Columbia: Workshop proceedings. BC Ministry of Forests, Research Branch, Victoria, BC Working Paper No. 57. Accessed January 2018 from: https://www.for.gov.bc.ca/hfd/pubs/docs/Wp/Wp57/Wp57-06.pdf .

Schiefer, E., Hassan, M. A., Menounos, B., Pelpola, C. P., & Slaymaker, O. (2010). Interdecadal patterns of total sediment yield from a montane catchment, southern Coast Mountains, British Columbia, Canada. *Geomorphology, 118*(1-2), 207-212.

Schoellhamer, D. H. (2002). Comparison of the basin-scale effect of dredging operations and natural estuarine processes on suspended sediment concentration. *Estuaries, 25*(3), 488-495.

Schubel, J. R., & Meade, R. H. (1977). Man's impact on estuarine sedimentation.

Schumann, E. H. (2015). Keurbooms Estuary floods and sedimentation. *South African Journal of Science, 111*(11-12), 1-9.

Seybold, H., Andrade, J. S., & Herrmann, H. J. (2007). Modeling river delta formation. *Proceedings of the National Academy of Sciences, 104*(43), 16804-16809.

Shaw, J., Stacey, C. D., Wu, Y., & Lintern, D. G. (2017). Anatomy of the Kitimat fiord system, British Columbia. *Geomorphology, 293*, 108-129.

Shaw, J, Conway K., and Kung R. (2018). *Distribution of hexactinellid sponge reefs in the chatham sound region, British Columbia*. Ottawa: Geological Survey of Canada

Simmons, S. M., Azpiroz-Zabala, M., Cartigny, M. J. B., Clare, M. A., Cooper, C., Parsons, D. R., ... & Talling, P. J. (2020). Novel acoustic method provides first detailed measurements of sediment concentration structure within submarine turbidity currents. *Journal of Geophysical Research: Oceans, 125*(5), e2019JC015904.

Simpson, M. R. (2001). *Discharge measurements using a broad-band acoustic Doppler current profiler* (p. 123). Reston: US Department of the Interior, US Geological Survey

Skeena Watershed Conservation Coalition (2014). *Pacific Northwest LNG Project – Public Comments re CEAA draft Environmental Assessment Report*. Retrieved 2018 from: https://www.ceaa.gc.ca/050/documents/p80032/108635E.pdf

St-Onge, G., & Hillaire-Marcel, C. (2001). Isotopic constraints of sedimentary inputs and organic carbon burial rates in the Saguenay Fjord, Quebec. *Marine Geology, 176*(1-4), 1-22.

Stommel, H. (1951). *Recent developments in the study of tidal estuaries* (No. WHOI-51-33). Woods Hole Oceanographic Institution, MA.

Svendsen, H., Beszczynska-Møller, A., Hagen, J. O., Lefauconnier, B., Tverberg, V., Gerland, S., ... & Azzolini, R. (2002). The physical environment of Kongsfjorden–Krossfjorden, an Arctic fjord system in Svalbard. *Polar research, 21*(1), 133-166.

Syvitski, J. P., Morehead, M. D., & Nicholson, M. (1998). HYDROTREND: a climate-driven hydrologic-transport model for predicting discharge and sediment load to lakes or oceans. *Computers & Geosciences, 24*(1), 51-68.

Syvitski, J. P., Peckham, S. D., Hilberman, R., & Mulder, T. (2003). Predicting the terrestrial flux of sediment to the global ocean: a planetary perspective. *Sedimentary Geology, 162*(1-2), 5-24.

Syvitski, J. P., & Milliman, J. D. (2007). Geology, geography, and humans battle for dominance over the delivery of fluvial sediment to the coastal ocean. *The Journal of Geology, 115*(1), 1-19.

Syvitski, J. P., Vörösmarty, C. J., Kettner, A. J., & Green, P. (2005). Impact of humans on the flux of terrestrial sediment to the global coastal ocean. *science*, *308*(5720), 376-380.

Syvitski, J. P., & Kettner, A. (2011). Sediment flux and the Anthropocene. *Philosophical Transactions of the Royal Society A: Mathematical, Physical and Engineering Sciences*, *369*(1938), 957-975.

Syvitski, J. P., Burrell, D. C., & Skei, J. M. (2012). *Fjords: processes and products*. Springer Science & Business Media.

Thompson, R. E. (1981). Oceanography of the British Columbia coast. *Can. Spec. Publ. Fish. Aquat. Sci.*, *50*, 1-291.

Thorne, P. D., & Hanes, D. M. (2002). A review of acoustic measurement of small-scale sediment processes. *Continental shelf research*, *22*(4), 603-632.

Topping, D. J., Wright, S. A., Melis, T. S., & Rubin, D. M. (2007, August). High-resolution measurements of suspended-sediment concentration and grain size in the Colorado River in Grand Canyon using a multi-frequency acoustic system. In *Proceedings of the 10th International Symposium on River Sedimentation* (Vol. 3, No. 470, p. 19). Moscow: World Association for Sediment and Erosion Research.

Trenhaile, A. S. (2010). *Geomorphology: a Canadian perspective*. 4th ed. Don Mills, Ont.: Oxford University Press.

Trites, R. W. (1956). The oceanography of Chatham Sound, British Columbia. *Journal of the Fisheries Board of Canada*, *13*(3), 385-434.

Turner, B., Nelson, J., Franklin, R., Weary, G., Hayward, B., Walker, T., McRae, C. (2007-10). *Geotour guide for Terrace, BC*. BC Geological Survey Geofile 2007-10.

Uncles, R. J. (1991). M4 tides in a macrotidal, vertically mixed estuary: the Bristol Channel and Severn. *Tidal Hydrodynamics*, *883*, 341-355.

Urick, R. J. (1948). The absorption of sound in suspensions of irregular particles. *The Journal of the acoustical society of America*, *20*(3), 283-289.

Valle-Levinson, A. (2011). 1.05 Classification of Estuarine Circulation. In *Treatise on estuarine and coastal science* (pp. 75-86). Academic Press Waltham.

Valle-Levinson, A. (2010). Definition and classification of estuaries. *Contemporary issues in estuarine physics*, *1*, 1-10.

Venditti, J. G., Church, M., Attard, M. E., & Haught, D. (2016). Use of ADCPs for suspended sediment transport monitoring: An empirical approach. *Water Resources Research*, *52*(4), 2715-2736.

Wagner, C. R., & Mueller, D. S. (2011). Comparison of bottom-track to global positioning system referenced discharges measured using an acoustic Doppler current profiler. *Journal of Hydrology*, *401*(3-4), 250-258.

Walters, C. J., Lichatowich, J. A., Peterman, R. M., & Reynolds, J. D. (2008). *Report of the Skeena Independent Science Review Panel to the Canadian Department of Fisheries and Oceans and the British Columbia Ministry of the Environment*, Victoria. British Columbia. Retrieved from: https://www.psf.ca/document-library/report-skeena-independent-science-review-panel

Wang, H., Yang, Z., Saito, Y., Liu, J. P., Sun, X., & Wang, Y. (2007). Stepwise decreases of the Huang He (Yellow River) sediment load (1950–2005): Impacts of climate change and human activities. *Global and Planetary Change, 57*(3-4), 331-354.

Webb, R., Rosenzweig, C.E., & Levine, E.R. (2000). Global Soil Texture and Derived Water-Holding Capacities. [Map data]. *ORNL DAAC*, Oak Ridge, Tennessee, USA. Retrieved 2018 from: http://dx.doi.org/10.3334/ORNLDAAC/548 (The water holding capacity layer was retrieved from: https://webmap.ornl.gov/ogcdown/wcsdown.jsp?dg_id=548_2)

Werner, 2016. GEN2 – GCM Selection and Statistical Downscaling Rationale. *PCIC*.

Winant, C. D., & Gutiérrez de Velasco, G. (2003). Tidal dynamics and residual circulation in a well-mixed inverse estuary. *Journal of Physical Oceanography, 33*(7), 1365-1379.

Wright, S. A., Topping, D. J., & Williams, C. A. (2010). Discriminating silt-and-clay from suspended-sand in rivers using side-looking acoustic profilers. In *Joint Federal Interagency Conference 2010: Hydrology and sedimentation for a changing future: existing and emerging issues*.

Wishart B. (2018). Building the First Canneries. Retrieved 2020 from https://heritageprincerupert.com/2018/12/05/building-the-first-cannery/

Wolanski & Elliott, E. (2007). *Estuarine ecohydrology*. Retrieved from https://ebookcentral.proquest.com

Yorke, T. H., & Oberg, K. A. (2002). Measuring river velocity and discharge with acoustic Doppler profilers. *Flow Measurement and Instrumentation, 13*(5-6), 191-195.

Chapter 1 Appendix

HydroTrend Inputs

14 Table 1Ch1Appx: HydroTrend Inputs

1981-2010 Variables	Subbasin 2 (SB2)	Subbasin 1 (SB1)	Whole Skeena Watershed (SW)	Method
Hypsometry File	N/A	N/A	N/A	The DEM for British Columbia was retrieved from the British Columbia Data Catalogue (GeoBC, 2012) and was published by the Ministry of Forests, Lands, Natural Resource Operations and Rural Development. The data was downloaded as individual map sheets at 30 meter resolution that were then merged in ArcGIS, projected in the Canada Albers Equal Area projection, and extracted to cover the Skeena subbasin areas defined by the National Hydro Network (NHN)-Geobase Series by Natural Resource Canada (NRCAN, 2016) polygons. After the elevation was categorized into elevation bins of 50 meters, the area within each of these elevation bins was calculated within ArcMap. These elevations were added together at each elevation step to produce the cumulative elevation at each fifty-meter step necessary for the Hydro.Hyps file. The Hypsometry file remains the same regardless of the climate scenarios.
Yearly mean annual Temperature (T) in $°C$ and total precipitation (P) in m; SD	ECCC Stations 5.8, 0, 1.2 0.9847, 0, 0.1496	ECCC Stations 7.05, 0, 0.95 1.98, 0, 0.2566	Same river mouth as SB1.	Annual temperature and precipitation at the Usk river mouth (SB2) or tidally drowned river mouth (SB1) was derived from an average of the nearest upstream and downstream ECCC climate station to the river mouth (Terrace and Hazelton for Usk) or from a mean of a 12.57 km area (2 km radius buffer) around the river mouth within a GCM raster during the climate scenario modelling. Instead of a starting value and change per year, a mean value over 30 years was calculated. This simplified the preparation of input files and was appropriate since the

				object is to compare changes over 30 year climate intervals rather than within the 10 yr. intervals.
Rain: Mass Balance Coefficient, Distribution Exponent, & Distribution Range	1.0, 1.4, 3	1.0, 1.2, 6	1.0, 1.3, 4	The rain mass balance coefficient is one way to account for a large basin wide difference in precipitation (CSDMS, 2018). Since the model application for the Skeena has been split into multiple subbasins where precipitation was averaged across the entire watershed at multiple elevations in an attempt to be representative of the entire basin, a rain mass balance coefficient of one was chosen for the model. A skewed Gaussian distribution is used within the HydroTrend model to create realistic tails for daily precipitation events (CSDMS, 2018). Daily precipitation is often right skewed due to a high frequency of low precipitation events and a low frequency of high precipitation events creating a long tail on the right hand side of the curve. The distribution exponent is the skewness factor between one (normal distribution) and two (highly skewed) of the distribution (CSDMS, 2018). Syvitski *et al.* (1998), mentioned that a distribution exponent of 1.3 was common for representing precipitation trends of many basins. Therefore, a rain mass balance exponent of 1.2 was applied to the coastal Skeena subbasins, and a slightly higher exponent of 1.4 was applied for the interior that would display more days without rainfall resulting in a mean distribution exponent of 1.3 for the entire Skeena watershed. The distribution range defines the width of the skewed Gaussian distribution and lies between 0 and 10 (CSDMS, 2018). The value between 0-10 is decided by the user where zero is a narrow curve with little variance between precipitation events, and 10 is a very wide distribution with high variance.

				When comparing the daily precipitation along the Skeena at Terrace, Hazelton, and Prince Rupert, the coastal areas receive more days with higher rainfall creating a wider distribution than those inland. Within the literature, a distribution range of 6 was applied to a coastal watershed of the Eel River in California (Syvitski *et al.*, 1998). Meanwhile, during a run of the HydroTrend model in the interior of Canada for the Mackenzie River, a distribution range of 1 was used (Lintern and Haaf, 2014). A distribution value of 6 was chosen for the coastal subbasin 1, 3 was chosen for the interior subbasin 2, and a mean value of 4 was chosen to represent the total watershed.
Constant annual base flow (m^3/s)	170	170	170	Derived from the mean minimum monthly flow value each year at Usk Hydrometric station (ECCC, 2020) averaged over the 1981-2010 period.
Monthly climate variables of mean T ($°C$), T SD, total monthly P (mm), and P SD.	ECCC Stations 1 -7.7, 4, 58.3, 47.3 2 -5, 3.2, 34.8, 20.5 3 -0.8, 2.1, 30.2, 9.1 4 4.1, 1.3, 29.7, 17.7 5 8.5, 1.8, 46.1, 18.7 6 12.3, 1.5, 63.5, 26.6 7 14.5, 1.1, 55.9, 23.8 8 14.0, 1.1, 49.5, 14.1 9 9.6, 1.4, 59.3, 31.5 10 3.9, 1.4, 70.8, 35.5 11 -2.7, 3.0, 64.8, 29.8 12 -6.8, 3.5, 55.3, 35.7	ECCC Stations 1 -0.3, 2.75 224.9, 70.6 2 0.9, 2,1 148.1, 74.8 3 3.3, 1.55, 146, 57.8 4 6.4, 1.15, 123.05, 52.9 5 9.8, 1.4, 97, 42.4 6 12.9, 1.1, 79.8, 35.7 7 14.95, 1.05, 85.75, 40.6 8 15.05, 1, 115.2, 51.4 9 11.8, 1, 188.9, 91.6 10 7.2, 1, 281.95, 82.4 11 2.5, 2.2, 252.05, 94 12 0.05, 2.4, 237.6, 110.4	ECCC Stations 1 -6.1, 3.7, 94.9, 52.5 2 -3.7, 2.9 59.8, 32.5 3 0.1, 2.0 55.7, 19.8 4 4.6, 1.3, 50.2, 25.4 5 8.8, 1.7, 57.3, 23.9 6 12.4, 1.4, 67.1, 28.6 7 14.6, 1.1, 62.4, 27.5 8 14.2, 1.1, 64, 22.3 9 10.1, 1.3, 87.8, 44.7 10 4.6, 1.3, 117.2, 45.8 11 -1.6, 2.8, 106, 43.9 12 -5.3, 3.3, 95.4, 52.1	**Historic:** Daily ECCC climate station variables (mean temperature and precipitation) were averaged monthly over stations with data available over the 1981-2010 period across the sub-basin. These values were as spatially spatial averaged across the watershed as possible. **Future:** For the 1980-2100 climate model runs under different RCPs, statistically downscaled climate scenarios raster grids for mean temperature maximum (TMax), temperature minimum (TMin), and precipitation (Pr) over the study area were downloaded from the Pacific Climate Impacts Consortium (PCIC, 2019) every five years. The climate scenarios do not take into account ENSO (PCIC, 2019). Therefore, averaging the data at a set interval over the period desired (every 5th year) would still be capturing the climate change trend without unintentionally selecting El Nino years. This was done to improve the

				computational efficiency in averaging the data for many variables, scenarios, and time intervals. The five-year increments were averaged into one raster grid representation of each 30 year interval (1981-2010, 2011-2040, 2041-2070, 2071-2100) and clipped over the sub-basin desired. From these averaged and clipped rasters, the mean Pr, TMax, TMin and associated standard deviations were calculated for each sub-basin, and TMax and TMin were used to calculate temperature mean.
Lapse rate to calculate freezing line ($°C/km$)	5.15	5.2	5.2	Syvitski *et al.* (2003), created a graph of lapse rate depending on the latitude that was used to determine the lapse rate based on the latitude at the river mouth for each subbasin.
Starting glacier ELA (m) and ELA change per year (m/a)	1898, 0.245	1605, 0.245	1741, 0.245	The mean glacier ELA for each subbasin was calculated in ArcMap using a dataset retrieved from the Rudolf Glacier Inventory (RGI, 2017) that was developed through an analysis between 2004-2006 on glaciers within British Columbia. Over the entire Skeena watershed area, a mean elevation of $1741m$ was calculated for glaciers logged within the RGI. A study on glacier mass balance in northern British Columbia and Alaska under different climate change scenarios (McGrath et al., 2017) was used to derive ELA change per year. Under RCP 6, ELA increases by consistently over 150 m within the northern Coast Mountains by the year 2100. Of the specific glacial ELA regressions available within the McGrath et al. (2017) paper, the Wolverine glacier was chosen opposed to a combination with interior glaciers because it was the only nearest glacial regression that remained above 150 m under the RCP 6 scenario (McGrath et al., 2017) as was described for the coast

				mountains. Since the regression is taken from further north than the Skeena, ELA change per year may be even higher than that estimated at the Wolverine Glacier. **Historic**: For the 1981-2010 period, only the last five years should have the change in glacial ELA applied since the starting ELA was calculated based on 2004-2006 (2005) levels. Thus, a glacial change per year for the 30 year 1981-2010 period is only ~0.245 m/annum based on the McGrath et al. (2017) RCP 4.5 scenario rate. **Future**: For the rest of the 30 year climate change scenarios from 2011 climate normal onwards, an ELA change of 1.47 m/annum under RCP 4.5 and 3.06 m/annum under RCP 8.5 were derived from total ELA regression between an 85 year period (2006-2015 compared to 1991-2100) for the coastal, Alaskan Wolverine Glacier (McGrath et al., 2017). RCP 2.6 was not discussed in the paper, so a value of half of the 4.5 RCP of 0.75 m/annum is applied. An ending ELA was also calculated at the end of each 30 year simulation in order to become the new starting ELA for the net climate normal.
Dry precipitation (nival and ice) evaporation fraction (ICE)	0.18	0.057	0.153	The dry precipitation (nival and ice) evaporation fraction falls between 0.0 and 0.9 and is used to estimate the percentage of snow and ice that will be evaporated (CSDMS, 2018). Cloud cover and days with precipitation would therefore decrease the dry precipitation evaporation fraction over a watershed. Lintern and Haaf (2014) have run HydroTrend over the Liard basin further north in British Columbia with an estimated dry precipitation evaporation fraction of 0.27. A sum of the days with precipitation (>=0.2 mm to >=25 mm) from Canadian Climate Normal

				resulted in a mean value of 146.1 days (~60% of the year without rain) with precipitation in a year for Laird basin over the 1981-2010 period (ECCC, 2019). At Hazelton, representing subbasin 2, had 219.3 days with precipitation (~40% of the year without precipitation) and at Terrace representing subbasin 1 had 334.2 days with precipitation (~0.084 % of the year). The dry precipitation evaporation fraction was then derived for the Skeena basins through a ratio in comparison to the Liard's DPEF and its percentage of the year without precipitation.
canopy interception alpha g (mm/d), beta g (-).	-0.1, 0.85	-0.1, 0.85	-0.1, 0.85	The model uses the canopy interception parameters to estimate how much precipitation is reaching the ground and contributing to runoff (CSDMS, 2018). The canopy interception values recommended on the CSDMS (2018) webpage of -0.1 and 0.85 was applied for the Skeena watershed.
groundwater pole evapotranspiration alpha_gwe (mm/d), beta_gwe (-).	10, 1	10, 1	10, 1	The model applies the groundwater pole evapotranspiration parameters to estimate the amount of water from the ground being taken up by plants and brought into the atmosphere by evapotranspiration (CSDMS, 2018). The recommended common values of 10 *mm/day* and 1 were used for the Skeena (CSDMS, 2018).
Delta plain gradient (m/m)	5.00e-4	5.00e-4	5.00e-4	The delta plain gradient in *m/m* is the average slope of the riverbed approaching the delta mouth. Delta plain gradient was derived from Gottesfeld and Rabnett (2008). Gottesfeld and Rabnett (2008) have stated the average gradient at the mouth of the river to be between 0.0-0.1%. They later state that the average gradient at Terrace is at 0.05% or approximately 0.4- 0.5 *m/km* (Gottesfeld and Rabnett, 2008).

Bedload rating term (-) (typically 1.0; if set to -9999, 1.0 will be used)	1.0	1.0	1.0	The bedload rating term was set to the recommended setting of 1.0 (CSDMS, 2018).
River basin length (km)	698.4	148.8	836	The River basin length for each subbasin was measured in ArcMap using Global Mapper satellite imagery and the NRCAN NHN (2016) rivers and streams layer.
Mean volume (km^3), (a)ltitude (m) or (d)rainage area of reservoirs (km^2)	0.007, a1082	0.0024 a867.5	0.0067 a974.75	The NRCAN NHN file (2016) contained all of the lakes and reservoirs within the Skeena watershed. Using the NRCAN (2016) NHN file along with the GeoBC (2012) DEM, the average altitude and area for all of the lakes and reservoirs were calculated in ArcMap. For the entire Skeena, a total of 1305737 lakes were listed within the Skeena watershed in the NHN file with a mean area of $0.67 \ km^2$ and mean altitude of 1058 m. Based on data from collected by Schiefer et al. (2001) and the Freshwater Fisheries Society of BC (2006) in lakes within the Skeena area, a mean depth of 10 m was chosen as reasonable for Skeena lakes.
River mouth velocity coef. (k) and exponent (m); v=kQ^m, River mouth width coef. (a) and exponent (b); w=aQ^b	0.844, 0.1 4.823, 0.5	0.557, 0.1 21.503, 0.5	0.557, 0.1 21.503, 0.5	Leopold & Maddock (1953) have revealed that common exponents at a river's mouth are 0.5, 0.1, and 0.4 for b, m, and f, respectively. River mouth velocity and width coefficient were calculated using channel width, depth, velocity, and discharge based on the hydraulic geometry formulas developed by Leopold & Maddock (1953). For SB2, mean discharge and water level were calculated at Usk Hydrometric station using data from ECCC (2020) and channel width was measured using Google Earth Satellite imagery. For SB1, at the start of the tidally drowned river mouth, the discharge was derived from BC Ministry of Environment

				(2020) discharge summation of multiple Skeena River tributaries. Also, at the start of the tidally drowned river mouth, depth was averaged across the channel from the nearest CHS multibeam bathymetry. Channel width was measured in ArcMap. Velocity was derived from the discharge divided by the width and depth according to the hydraulic geometry formula (Leopold & Maddock, 1953).
Average river velocity (m/s)	1.668	1.668	1.668	Velocity was derived from the 1981-2010 mean discharge at Usk hydrometric station divided by the mean water level from Usk and mean channel width measured in ArcMap using Digital Globe satellite imagery. Since it was the only long-term discharge station where velocity could be calculated, and it was centrally located within the watershed, the velocity at Usk was used as the average for the river for all subbasins.
Maximum/minimum groundwater storage (m^3)	1.6e+10 1.37e+10	5.57e+09 4.92e+09	2.16e+10 1.86e+10	A global data set by Webb *et al.* (2000) from Oak Ridge National Laboratory Distributed Active Archive Center (ORNL DAAC), which displays a raster of estimates of global soil texture and derived water holding capacity across the globe per arc second grid blocks, was used to estimate the minimum and maximum groundwater stored within the Skeena. Within ArcGIS, the minimum and maximum storage for each pixel type was multiplied over each pixel area and added together. Lintern and Haaf (2014) also used the Web et al. (2000) water holding capacity dataset to estimate groundwater storage inputs for HydroTrend over the Mackenzie basin.
Initial groundwater storage (m^3)	1.485e+10	5.245 e+09	2.01 e+10	The initial groundwater storage was set to the mean of the minimum and maximum groundwater storage to reduce the runup time of the model.

Ground water (subsurface storm flow) coefficient (m^3/s) and exponent (unitless)	18142.8, 1.0	4301.0167 1.0	19420 1.0	SSF coefficient and exponent required by the model, the coefficient and exponent were adapted from those used by Linter and Haaf, (2014) over the Liard basin. Fortunately, the Liard basin is of a relatively similar surficial material as the Skeena. However, the Skeena and Liard basins are very different sizes. Therefore, the SSF and total area for the Liard basin from Lintern and Haaf (2014) was scaled to match the area of the Skeena subbasins. The SSF exponent was set to one, which was the exponent used within all subbasins of the Mackenzie by Lintern and Haaf (2014).
Saturated hydraulic conductivity (mm/day)	77.11	36.55	68.19	Proportions of surface lithology for the Skeena were estimated from (Fulton, 1995) for each subbasin (for subbasin 2, estimations of surface ground cover are approximately 40% bedrock outcroppings, 45% glacial till, 10% mixed alluvium and fine-grained lake deposits, and 5% marine sand). Applied to the soil textural and related saturated hydraulic conductivity chart shared on the CSDMS website (2018), a hydraulic conductivity value of 121.91 *mm/day* was used to describe the glacial till substrate. A medium texture of silt with a hydraulic conductivity of 36. 55 *mm/day* was chosen to represent alluvium. The marine sand and complex material were attributed a loam sand- sandy loam texture of 364.95 *mm/day*. Based on the CSDMS hydraulic conductivity table, a rough estimate of 1 *mm/day* was applied to the bedrock. A weighted average based on the proportion of the subbasin covered by each landcover type was used to compute the total average hydraulic conductivity for the sub-basin.
Longitude, latitude position of	-128.43, 54.63	-130.124 54.132	-130.124 54.132	Latitude and longitude retrieved from ArcMap.

the river mouth (decimal degrees)				
BQART Equation: Lithology factor from hard - weak material (0.3 - 3)	2	1	1.5	A lithology factor was chosen based on the classification scheme defined by Syvitski and Milliman (2007). According to Syvitski and Milliman (2007), a lithology factor of 1 is intended for areas consisting of volcanic rock or a mixture of hard to soft lithologies. A lithology factor of 2 represents an area which consists of a greater proportion of glacial till and clastic sediments. A lithology factor of 1.5 represents softer-mixed lithology (Syvitski and Milliman (2007).
BQART Equation: Anthropogenic factor (0.5 - 8), human disturbance of the landscape	1	1	1	Syvitski and Milliman (2007) have defined the anthropogenic factor on a global scale based on population density and gross national product (GNP) per capita. For basins around dense cities in the United States and Europe, an E_h factor of 0.5 is recommended due to a high population density, GNP/capita, and human influence on soil erosion. An E_h factor of one was displayed for most of the globe and was described as areas with a low human footprint or a mixture of soil erosion and conservation drivers. Basins in parts of Asia, with a high population, but low *GNP/capita* or those at their historical peak of forestation of open pit mining are recommended with an E_h value of 2.0 (Syvitski and Milliman, 2007). Although the Skeena is influenced by forestry, the impacts appear lower than those in other basins on a global scale. Therefore, an E_h factor of 1 was chosen for the Skeena watershed.
Other				The proportion of suspended sediment in each size class was left at 1 for each model simulation as this is a proportional calculation that can

				easily be applied after the model is run and using the final result simulations. Only one outlet at the delta was applied as this is another proportional calculation that can be applied later after simulating the total output at the river mouth. Zero sediments are assumed to be trapped in the delta as the goal of the HydroTrend simulation is to address the total output of the Skeena River to potentially build a delta or feed the estuary. Delta deposition and trapping will be addressed at the SedFlux modelling stage.

HydroTrend Input Uncertainty Analysis

Over the 1991-2100 period, McGrath et al. (2017) estimated an ELA change of 1.47 m/annum under RCP 4.5 and 3.06 m/annum under RCP 8.5 for the Wolverine Glacier in the Alaskan Coast Mountains. Glacier ELA for the Skeena region is derived from the 2004-2006 Rudolf Glacier inventory. Since the ELA was based on ~2005, the last ~5 years of the 1981-2010 historical HydroTrend simulations need a glacier change per year applied from the RCP 8.5 or 4.5 scenario. This period of change for the last five years works out to 0.245 m/annum for RCP 4.5 and 0.51 m/annum for RCP 8.5, using estimates by McGrath et al. (2017). Although the difference is small, a higher glacial change rate increases the modelled discharge and produces a higher mean percent difference compared to measurements at Usk station (Table 2CH1Appx). Therefore, the lower 0.245 m/annum for RCP 4.5 was applied for glacial change over all the historic 1981-2010 simulations.

Lintern and Haaf (2014) used an ICE of 0.27 over the Liard basin in the interior of northern British Columbia. However, the Skeena watershed receives a considerably higher number of precipitation days and cloud cover, which could reduce the evaporation of ice and snow. Applying the 0.27 value to the Skeena SB2 produces lower discharge and a higher mean difference from Usk Station due to an increase in evaporation across the basin (Table 2CH1Appx). A lower value of 0.18 produced more accurate results (Table 2CH1Appx) and was selected through a comparison of the number of days with precipitation in Hazelton to that of a central ECCC station in the Liard Basin. Therefore, the lower ICE value is applied for the 1981-

2010 period. The selection of the 0.18 value is also discussed in further detail in Table 1Ch1Appx.

15 Table 2Ch1Appx: Uncertainty Analysis Results.
1981-2010 HydroTrend (HT) simulations were averaged into 12 monthly values to compare to Usk Hydrometric mean monthly discharge (Q).

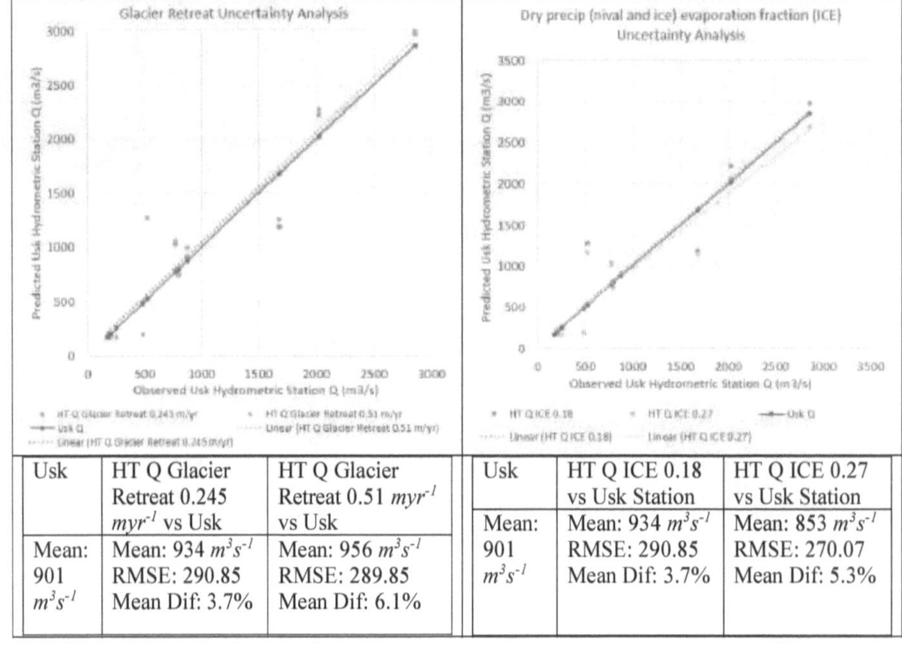

Usk	HT Q Glacier Retreat 0.245 myr^{-1} vs Usk	HT Q Glacier Retreat 0.51 myr^{-1} vs Usk
Mean: 901 m^3s^{-1}	Mean: 934 m^3s^{-1} RMSE: 290.85 Mean Dif: 3.7%	Mean: 956 m^3s^{-1} RMSE: 289.85 Mean Dif: 6.1%

Usk	HT Q ICE 0.18 vs Usk Station	HT Q ICE 0.27 vs Usk Station
Mean: 901 m^3s^{-1}	Mean: 934 m^3s^{-1} RMSE: 290.85 Mean Dif: 3.7%	Mean: 853 m^3s^{-1} RMSE: 270.07 Mean Dif: 5.3%

Current Radiative Forcing

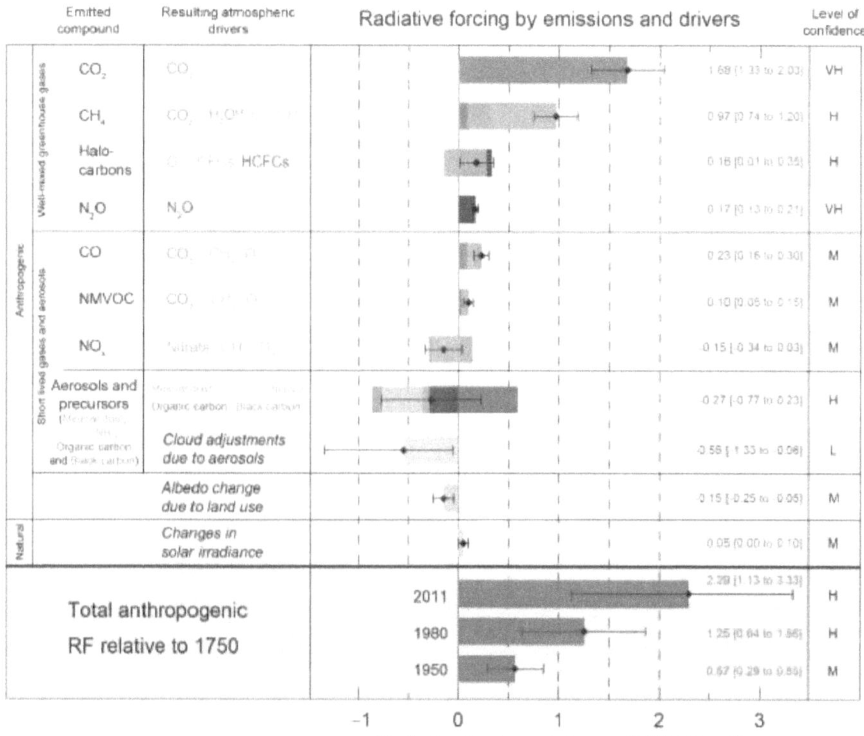

43 Figure 1Ch1Appx: 1950-2011 radiative forcing compared to pre-industrialization (1750). *Retrieved and unmodified from IPCC (2013).*

Chapter 2 Appendix

Radiocarbon Dated Cores

16 Table 1Ch2Appx: Radiocarbon Dated Core Uncalibrated and Calendar Ages with Sedimentation Rates

Estuary Location Core Number	Core Stratigraphy	Sample Name	Core Depth (*cm*)	Beta analytic lab. #	Dated Material	Conventional radiocarbon age (BP) ± SD	Mean Calendar Age (BP) ± SD	Mean Grain size (PHI)	Sedimentation Rate (SR) (*cmyr⁻¹*) (1 Sig Fig)	Surface Sedimentation Rate Category (SedR)*
Ogden Channel										
2016002 122	Single Silt Unit	122a	0-2							2>SR>1
		122b	108	186248	shell	920 ± 15	143 ± 72		122a - 122b: 0.8	
		122c	221	186249	shell	995 ± 15	218 ± 72		122b - 122c: 2	
2016002 123	Single Silt Unit	123a	0-2					6.46		3>SR>2
		123b	131	186250	shell	830 ± 15	77 ± 59		123a - 123b: 2	
		123c	489	186251	shell	970 ± 15	189 ± 72		123b - 123c: 3	
2016002 124	Single Silt Unit	124a	0-2					6.7		2>SR>1
		124b	181	186252	shell	880 ± 15	110 ± 70		124a - 124b: 2	
2016002 129	Many Varied Interbedde	129a	0-2					5.58		NA
		129b	44	186274	wood	120 ± 15	131 ± 76		129a - 129b: 0.3	

	d Coarse & Fine Units	129c	160					2.87		
		129d	226	18627 5	wood	155 ± 15	161 ± 83	2.88	129b - 129d: 6	
			242					2.73		
201600 2 134	Many Varied Interbedde d Coarse & Fine Units	134a	0-2					3.57		NA
		134b	165					4.48		
		134c	242	18627 6	wood	340 ± 15	389 ± 47	3.09	134a - 134c: 0.6	
		134d	354	18627 7	wood	800 ± 15	710 ± 13	3.32	134c - 134d: 0.3	

Arthur Passage

201500 2 147	2-4 Different Units	147a	0-2					4.22		1>SR>0.1
		147b	51	16751 4	shell	925 ± 15	148 ± 72		147a - 147b: 0.3	
		147c	183	16751 5	shell	1255 ± 15	475 ± 47	6.32	147b - 147c: 0.4	
		147d	249	16751 6	Shell	1925 ± 20	1080 ± 67	5.05	147c - 147d: 0.1	
201500 2 137	2-4 Different Units	137a	0-2					4.09		0.1>SR>0
		137b	81	18625 3	shell	1435 ± 15	603 ± 43	4.75	137a - 137b: 0.1	
		137c	106	18625 4	shell	2125 ± 15	1277 ± 52	4.78	137b - 127c: 0.04	
		137d	126	18625 5	Shell	2955 ± 15	2220 ± 66		137c - 137d: 0.02	
		137e	145					5.16		
		137f	194					7.39		

Proximal - Base Sands

201600 2 138	Single Silt Unit Single Silt Unit	138a	0-2					5.47		2>SR>1
		138b	140	186257	shell	895 ± 15	122 ± 72		138a - 138b: 1	
201600 2 139	Single Silt Unit	139a	0-2					5.32		2>SR>1
		139b	225	186278	wood	130 ± 15	135 ± 80		139a - 139b: 2	
201600 2 162	Single Silt Unit	162a	0-2					6.1		2>SR>1
		162b	40	186259	shell	890 ± 15	118 ± 71		162a - 162b: 0.3	
		162c	127	186260	shell	940 ± 15	161 ± 72		162b - 162c: 2	
201600 2 161	Many Varied Interbedded Coarse & Fine Units	161a	0-2					6.49		NA
		161b	6	186258	shell	135 ± 15	35 ± 9		161a - 161b: 0.2	
		161c	63					1.51		
		161d	91					2.68		
		161e	116	186280	wood	175 ± 15	168 ± 86		161b - 161e: 0.8	
		161f	249					2.77		
		161g	271	186281	wood	375 ± 15	435 ± 57		161e - 161g: 0.6	
Inner Chatham Sound										
201600 2 164	Single Silt Unit	164a	0-2					5.96		2>SR>1
		164b	178	186262	shell	825 ± 15	75 ± 57		164a - 164b: 2	
		164c	217	186263	shell	910 ± 15	135 ± 73		164b - 164c: 0.7	

		164d	350	186264	shell	1045 ± 15	294 ± 70		164c - 164d: 0.8	
201600 2 165	Single Silt Unit	165a	0-2					6.37		1>SR>0.1
		165b	157	186265	shell	965 ± 15	184 ± 71		165a - 165b: 0.9	
		165c	268	186266	shell	1175 ± 15	404 ± 54		165b - 165c: 0.5	
		165d	443	186267	shell	1610 ± 15	760 ± 58		165c - 165d: 0.5	
201600 2 166	Single Silt Unit	166a	0-2					6.33		1>SR>0.1
		166b	91	186282	wood	130 ± 15	135 ± 80		166a - 166b: 0.7	
201600 2 167	Single Silt Unit	167a	0-2					6.83		1>SR>0.1
		167b	159	186268	shell	1175 ± 15	404 ± 54		167a - 167b: 0.4	

Outer Chatham Sound

201500 2 148	2-4 Different Units	148a	0-2					4.54		0.1>SR>0
		148b	38	167517	shell	9300 ± 20	9561 ± 65	4.12	148a - 148b: 0.004	
		148c	268	167518	shell	10870 ± 25	11611 ± 156	7.17	148b - 148c: 0.1	
		148d	386	167519	shell	12205 ± 30	13279 ± 64		148c - 148d: 0.1	
201500 2 149	2-4 Different Units	149a	0-2					6.15		0.1>SR>0
		149b	40	167520	shell	4160 ± 15	3696 ± 75	6.23	149a - 149b: 0.01	
		149c	94	167521	shell	10340 ± 25	10907 ± 99	6.57	149b - 149c: 0.01	

		149d	246	16752 2	shell	10580 ± 25	11174 ± 61		149c - 149d: 0.6	
		149e	294	16752 3	shell	11330 ± 30	12460 ± 103		149d - 149e: 0.04	
201500 2 150	2-4 Different Units	150a	0-2					6.59		1>SR>0.1
		150b	150	16752 4	shell	1805 ± 15	956 ± 62		150a - 150b: 0.2	
		150c	300	16752 5	shell	10730 ± 25	11381 ± 130		150b - 150c: 0.01	
		150d	451	16752 6	shell	11170 ± 30	12233 ± 131	6.04	150c - 150d: 0.2	
		150e	488	16752 9	shell	12355 ± 35	13404 ± 68	4.49	150d - 150e: 0.03	

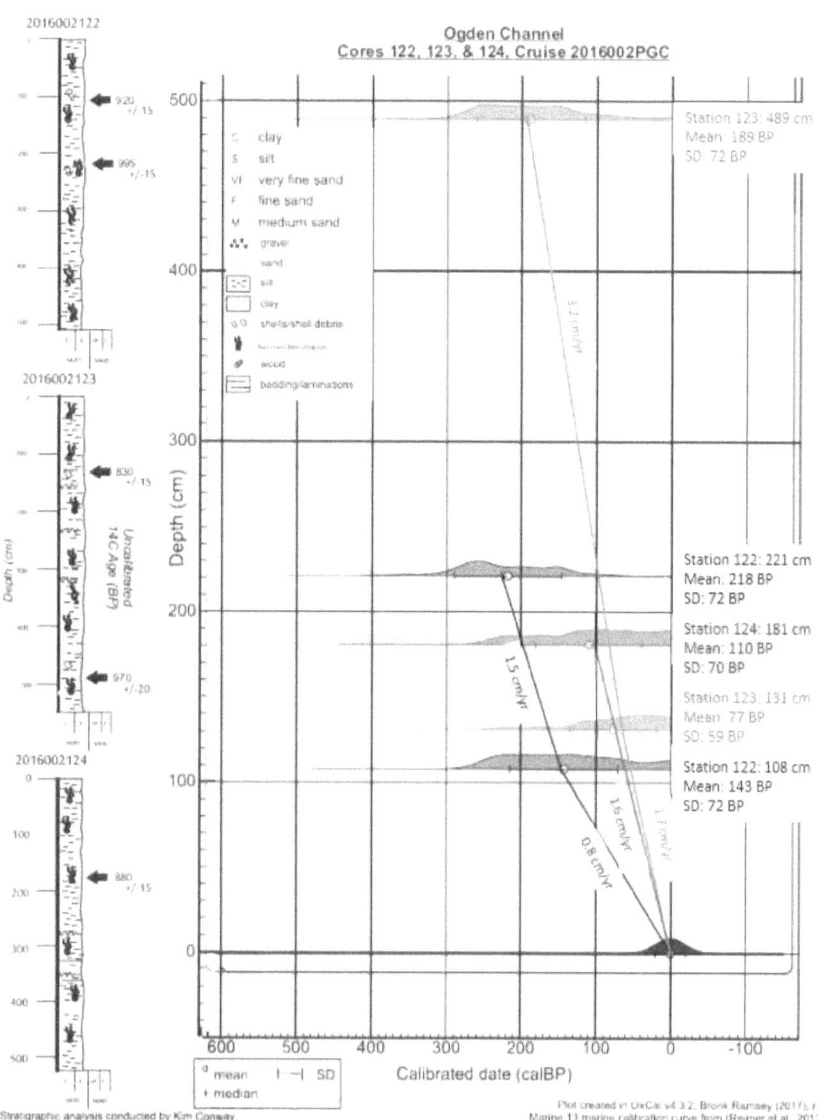

44 *Figure 1Ch2Appx: Ogden Channel Sedimentation Rate Diagrams.*

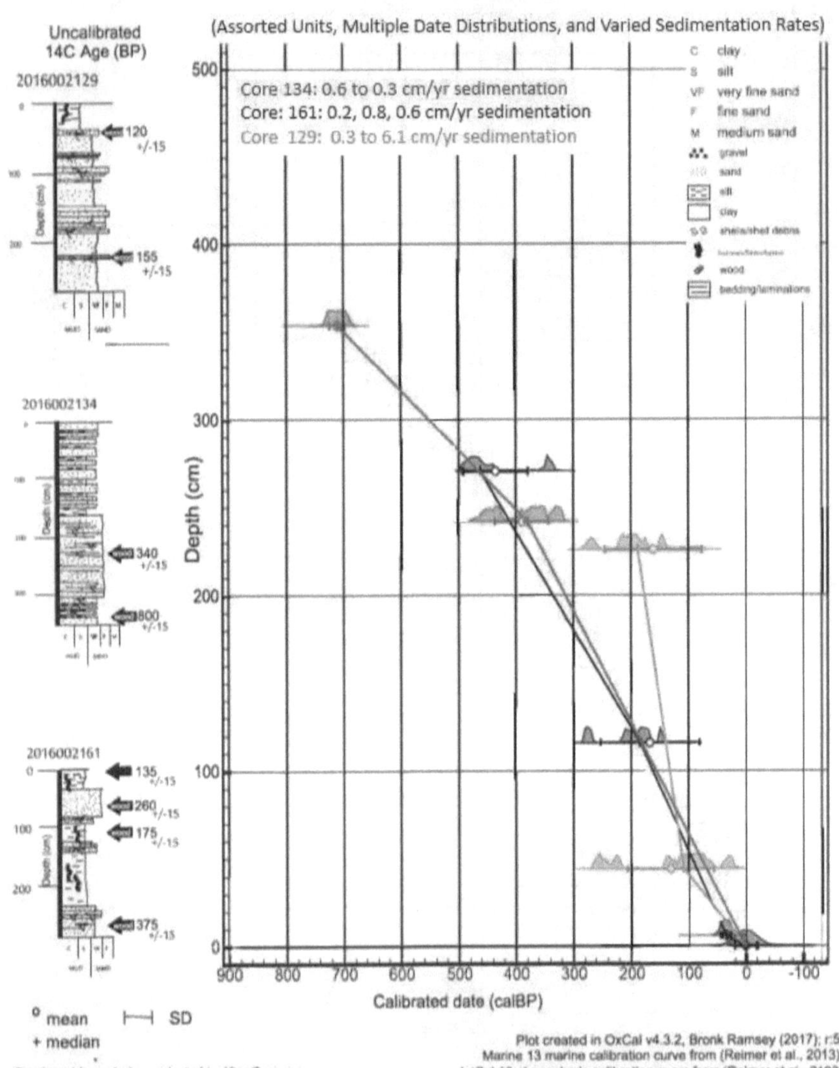

45 Figure 2Ch2Appx: Subaqueous Channel Sedimentation Rate Diagrams

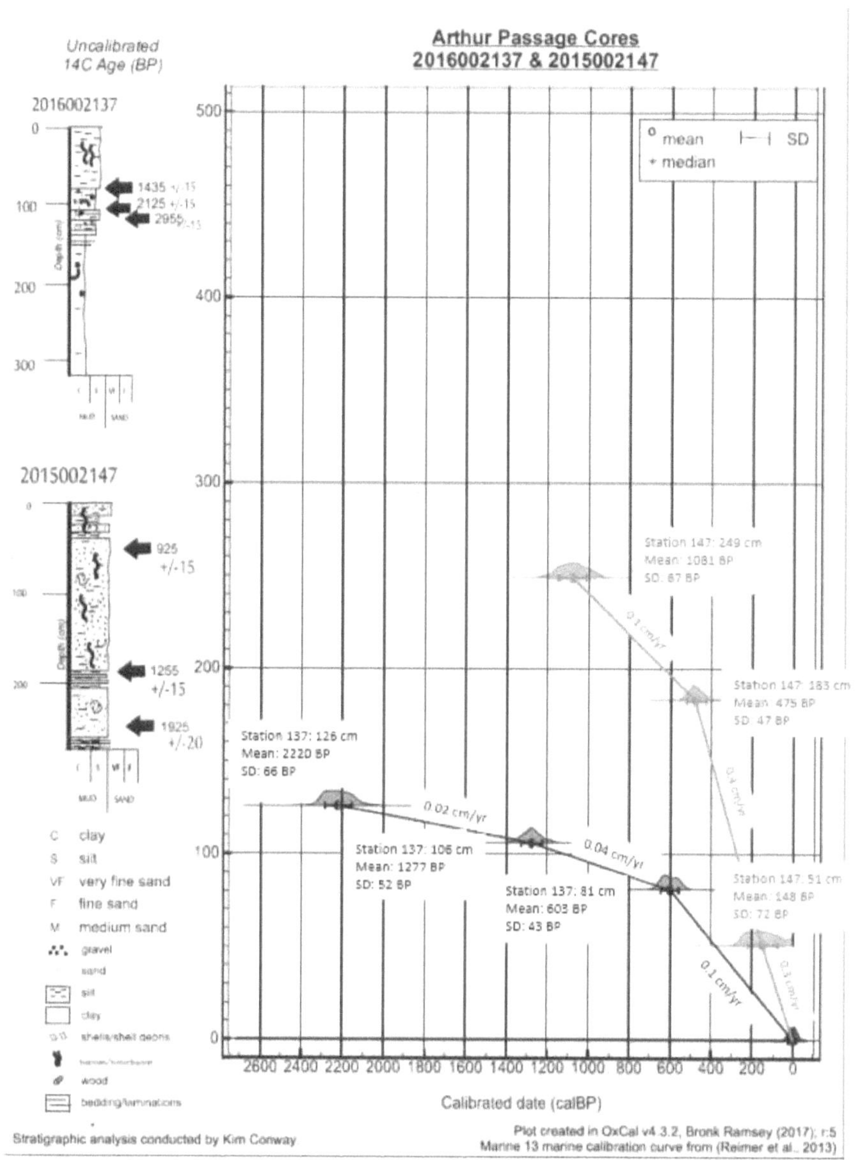

46 Figure 3Ch2Appx: Arthur Passage Sedimentation Rate Diagrams.

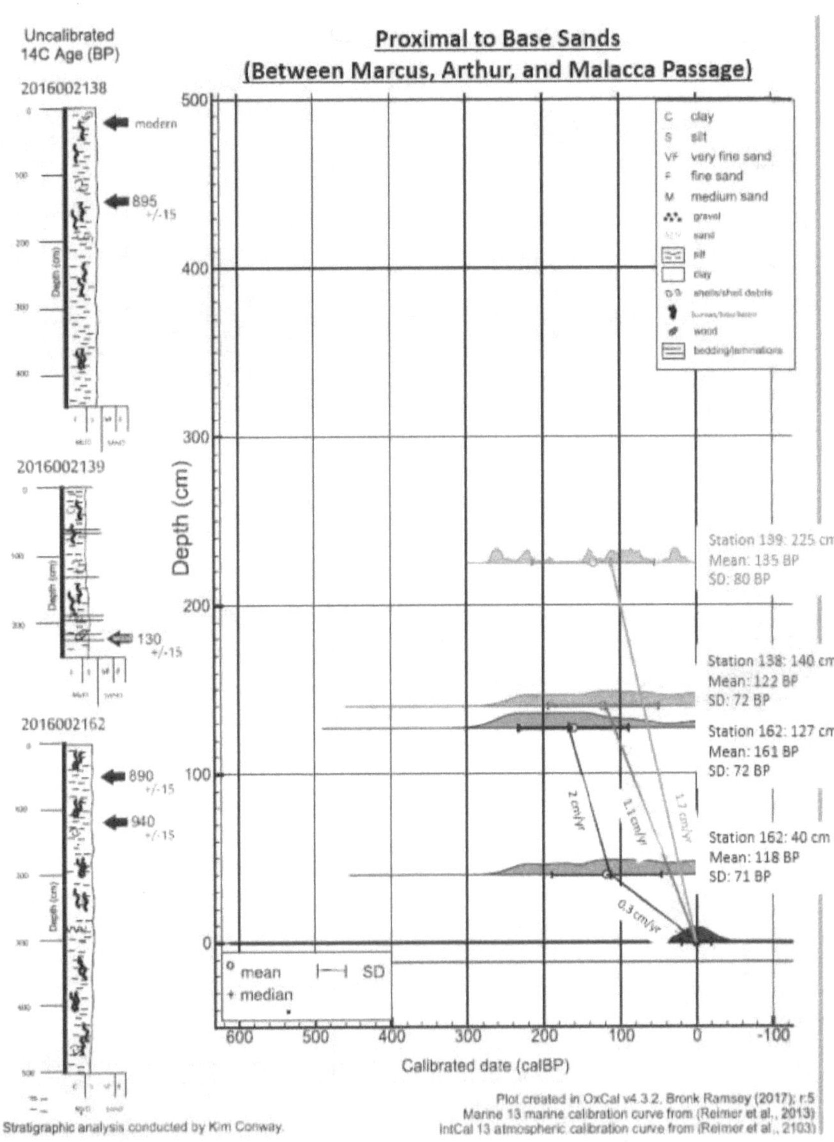

47 *Figure 4Ch2Appx: Proximal to Base Sands Sedimentation Rate Diagrams.*

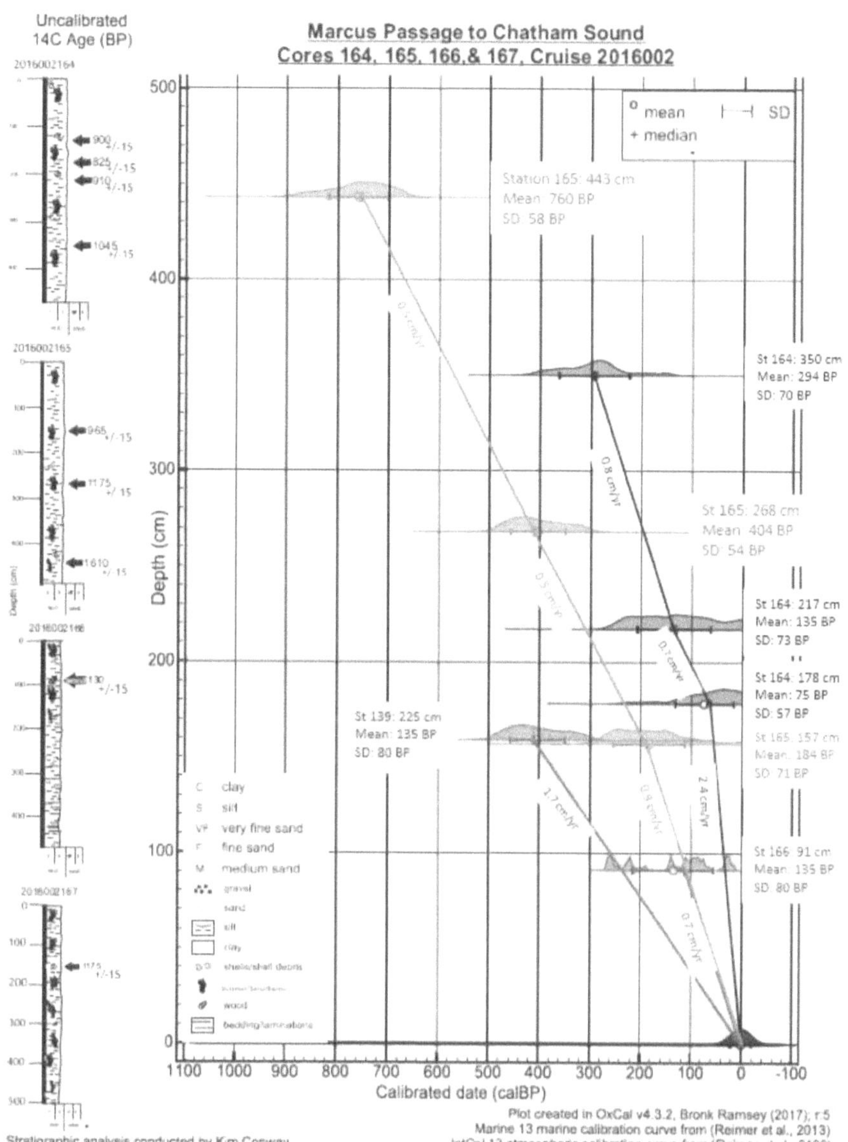

48 Figure 5Ch2Appx: Marcus Passage Sedimentation Rate Diagrams

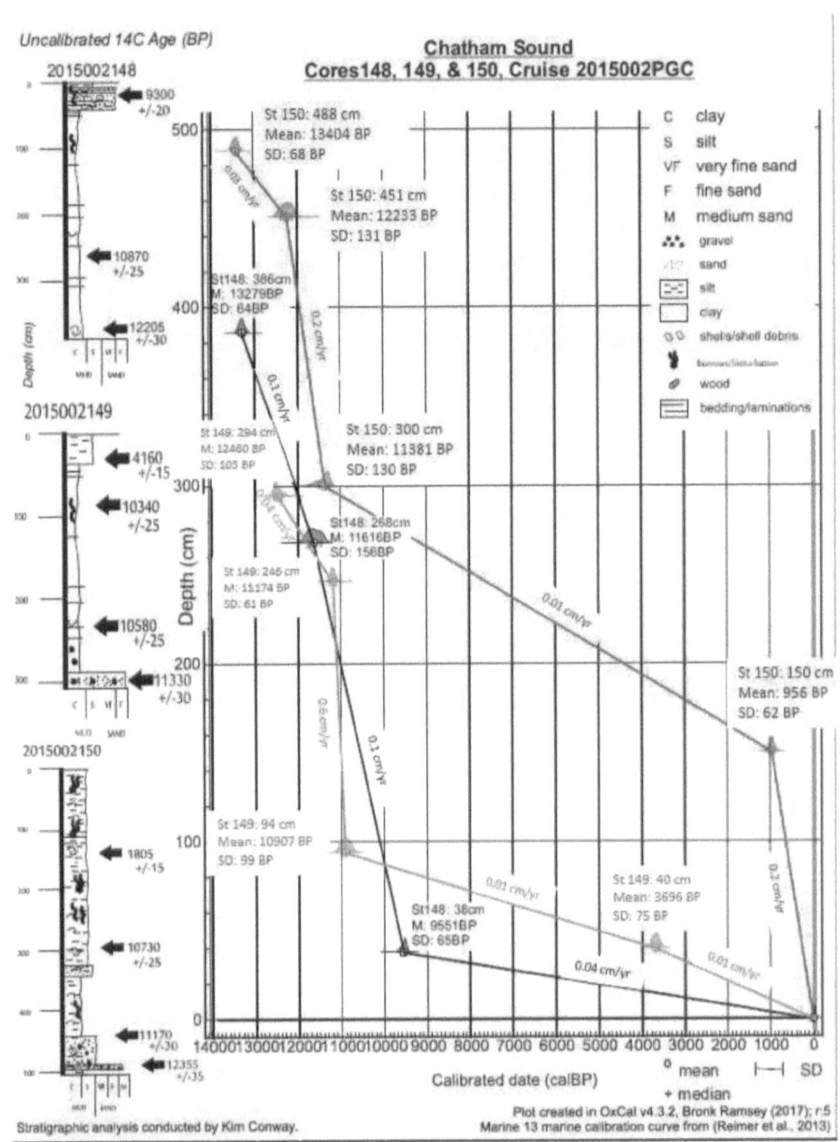

49 Figure 6Ch2Appx: Chatham Sound Sedimentation Rate Diagrams.

Chapter 3 Appendix

ADCP, CTDTu, and Water Sample Survey Details

17 Table 1Ch3Appx: ADCP, CTDTu, and Water Sample Transects

Date	Location	Usk Discharge (m3/s)	Tidal Cycle Surrounds High or Low Water	Survey Notes
5/2/2019	Skeena Downstream Ecstall (SDE)	610	High water	Ebb, flood, and slack ADCP surveyed continously with CTD and water samples (WS) to depth at multiple points across the channel. Suspended sediment grain size samples were obtained in addition to the water sample total suspeneded, inorganic, and organic material concentrations collected in all other passages.
5/2/2019	Inverness Passage (IV)	610	High water	Only flood tide ADCP surveyed with water samples and CTD casts to depth in the channel center. No ebb tide ADCP transect leaving only the CTD and surface WS as functional for the ebb. Inverness sampling occurred during the periphery of SDE surveying.
5/3/2019	Telegraph Passage (TP)	690	High water	Ebb, flood, and slack ADCP surveyed continously with CTD and water samples to depth at multiple points across the channel.
5/1/2019	Marcus Passage (M)	580	High water	Ebb, flood, and slack ADCP surveyed continously with CTD and water samples to depth at multiple points across the channel. One ADCP transect during the slack tide was lost due to a power outage.
8/20/2019	Skeena Downstream Ecstall (SDE)	1260	Low water	Ebb, flood, and slack ADCP surveyed continously with CTD and water samples to depth at multiple points across the channel.
8/19/2019	Inverness Passage (IV)	1040	Low water	Ebb and flood ADCP surveyed continously with CTD and water samples to depth at multiple points across the channel. Slack tide was surveyed further landward in the channel than ebb and flood with no CTD or water samples casts.
8/22/2019	Telegraph Passage (TP)	2020	Low water	Ebb, flood, and slack ADCP surveyed continously with CTD and water samples to depth at multiple points across the channel.
8/23/2019	Skeena Upstream Ecstall (SUE)	2080	Low water	Ebb and slack surveyed continously with CTD and water samples to depth at multiple points across the channel. Flood tide was not surveyed.

Hydraulic Geometry of Skeena Estuary Passages

18 Table 2Ch3Appx: Hydraulic Geometry calculations of select Skeena Passages

Location	Mean Width (m)	Mean Depth (m)	Mean Velocity (m/s)	Mean Tidal Passage Volume (m3/s)
Skeena Downstream Ecstall	3500	11.87	0.65	26997
Inverness Passage (IV)	350	14.06	0.65	3199
Telegraph Passage (TP)	2450	16.04	0.65	25549
Marcus Passage (M)	2700	14.31	0.65	25117
Skeena Upstream Ecstall (SU	2150	6.00	0.65	8385

Velocity over Time of ADCP Surveys over a Tidal Cycle

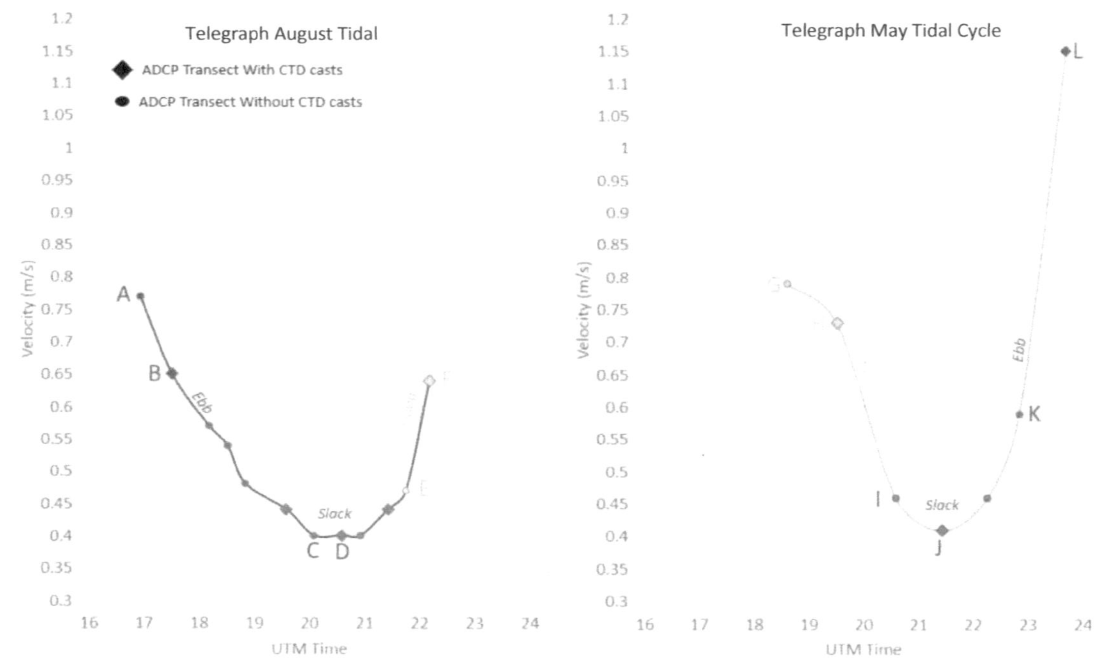

50 Figure 1CH3Appx: T Mean tidal cycle UTC time and velocity of ADCP cross-channel transects. ADCP transects plotted on Figures 7Ch3, and 8Ch3 are labelled A-L. All ADCP transects taken are displayed as a circle (ADCP only) or diamond (ADCP transect with CTDTu casts). Based on a visual analysis of the flow direction within ADCP transects, purple is used to represent ebbing conditions, green for slack, and yellow for the flood.

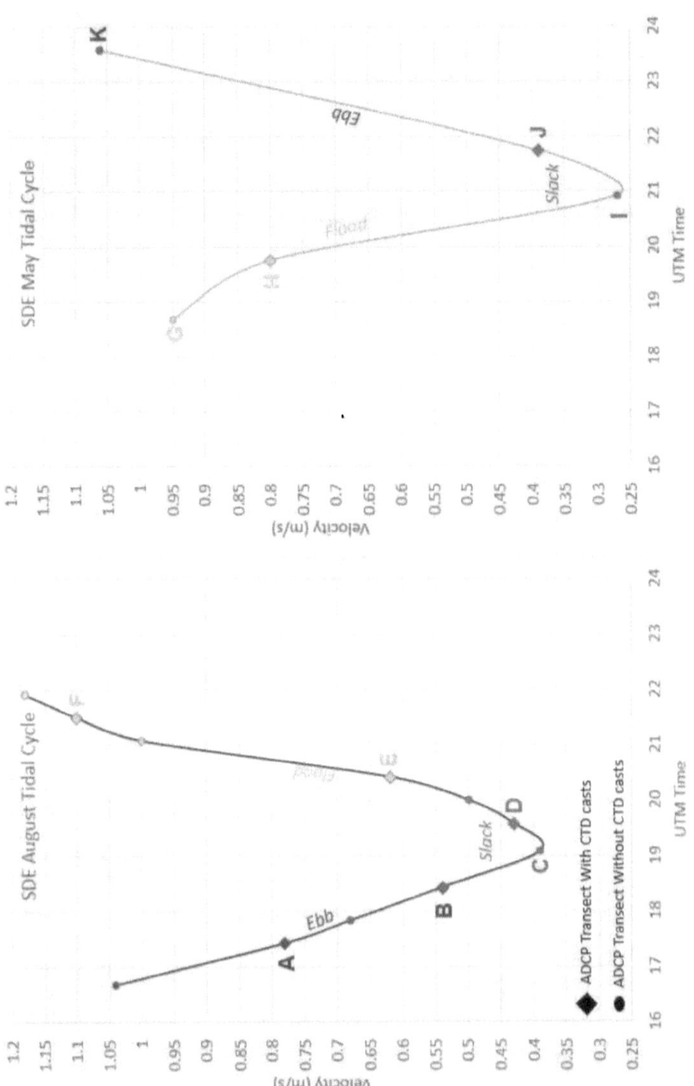

51 *Figure 2Ch3Appx: SDE Mean tidal cycle UTC time and velocity of ADCP cross-channel transects. ADCP transects plotted on Figures 9Ch3, and 10Ch3 are labelled A-K. All ADCP transects taken are displayed as a circle (ADCP only) or diamond (ADCP transect with CTDTu casts). Based on a visual analysis of the flow direction within ADCP transects, purple is used to represent ebbing conditions, green for slack, and yellow for the flood.*